AQA Chemistry

A LEVEL
YEAR 1
AND AS

Emma Poole

OXFORD
UNIVERSITY PRESS

Great Clarendon Street, Oxford, OX2 6DP, United Kingdom

Oxford University Press is a department of the University of Oxford.
It furthers the University's objective of excellence in research,
scholarship, and education by publishing worldwide. Oxford is a
registered trade mark of Oxford University Press in the UK and in
certain other countries

British Library Cataloguing in Publication Data
Data available

978-0-19-835183-2

10 9 8 7 6 5 4 3 2 1

Paper used in the production of this book is a natural, recyclable
product made from wood grown in sustainable forests.
The manufacturing process conforms to the environmental regulations
of the country of origin.

Printed in Great Britain by Bell and Bain Ltd. Glasgow

Artwork by Q2A Media

AS/A Level course structure

This book has been written to support students studying for AQA AS Chemistry and for students in their first year of studying for AQA A Level Chemistry. It covers the AS sections from the specification, the content of which will also be examined at A Level. The sections covered are shown in the contents list, which also shows you the page numbers for the main topics within each section. If you are studying for AS Chemistry, you will only need to know the content in the blue box.

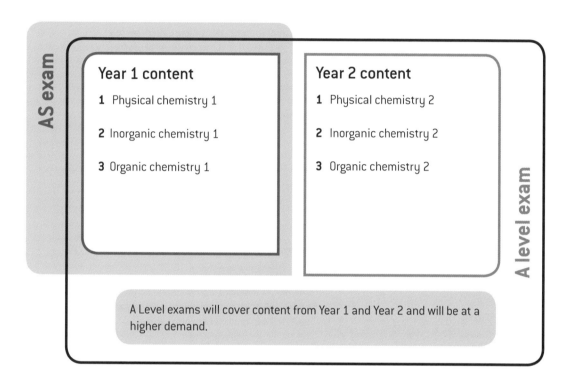

AS exam

Year 1 content

1 Physical chemistry 1

2 Inorganic chemistry 1

3 Organic chemistry 1

Year 2 content

1 Physical chemistry 2

2 Inorganic chemistry 2

3 Organic chemistry 2

A level exam

A Level exams will cover content from Year 1 and Year 2 and will be at a higher demand.

Contents

How to use this book

This book contains many different features. Each feature is designed to foster and stimulate your interest in chemistry, as well as supporting and developing the skills you will need for your examinations.

Worked example

Step-by-step worked solutions.

Summary questions

1 These are short questions at the end of each topic.

2 They test your understanding of the topic and allow you to apply the knowledge and skills you have acquired.

3 The questions are ramped in order of difficulty. Lower-demand questions have a paler background, with the higher-demand questions having a darker background. Try to attempt every question you can, to help you achieve your best in the exams.

Shows you what you need to know for the practical aspects of the exam.

Common misconception

Common student misunderstandings clarified.

Go Further

To push you a little further.

Specification references

→ At the beginning of each topic, there are specification reference to allow you to monitor your progress.

Key term

Pulls out key terms for quick reference.

Synoptic link

These highlight how the sections relate to each other. Linking different areas of chemistry together becomes increasingly important, and you will need to be able to do this.

Revision tip

Prompts to help you with your revision.

Maths skill

A focus on maths skills.

Chapter 1 Practice questions

Practice questions at the end of each chapter and each section, including questions that cover practical and maths skills.

1 Use the Periodic Table to deduce the full electron configuration of: *(2 marks)*

   a Mg

   b Na^+

2 A sample of magnesium was analysed and found to contain three isotopes. The percentage abundance of each isotope is shown below.

Isotope	Percentage abundance / %
^{24}Mg	78.0
^{25}Mg	10.0
^{26}Mg	12.0

Use the information in the table to calculate the relative atomic mass of this sample of magnesium. Give your answer to three significant figures. *(2 marks)*

3 Complete the table to show the relative charge and relative mass of these sub-atomic particles. *(3 marks)*

Sub-atomic particle	Relative charge	Relative mass
proton		
neutron		

Key term

Atomic number (Z): The number of protons in the nucleus.

Key term

Mass number (A): The number of protons and neutrons in the nucleus.

Key term

Isotope: The isotopes of an atom of an element contain different numbers of neutrons but the same number of protons and also of electrons. So they have the same atomic number but a different mass number.

Summary questions

1 Write down the numbers of protons, neutrons, and electrons in the following:

   a $_{11}^{23}\text{Na}$

   b $_{8}^{16}\text{O}^{2-}$

   c $_{35}^{80}\text{Br}$

   d $_{19}^{40}\text{K}^{+}$ (4 marks)

2 Explain why chlorine-35 and chlorine-37 are chemically identical.
                        (1 mark)

3 Describe the relative mass and relative charge of the sub-atomic particles found in the nucleus of an atom.   (2 marks)

Atoms and sub-atomic particles

Atoms are made from sub-atomic or fundamental particles called protons, neutrons, and electrons. The masses and charges of all these fundamental particles are very small, so we always look at them relative to the mass and charge of a proton.

Particle	Relative mass	Relative charge
proton	1	+1
neutron	1	0
electron	$\dfrac{1}{1836}$	−1

The mass of an electron is so small it is often considered to be negligible. Protons and neutrons are located in the nucleus of the atom while electrons orbit the nucleus.

Atomic number and mass number

Ensure that you can recall the definitions of atomic number and mass number.

- Atoms can be represented: $_{Z}^{A}\text{X}$
- Number of protons = atomic number (Z)
- Number of electrons = atomic number (Z) (ONLY for neutral atoms)
- Number of neutrons = mass number − atomic number (A − Z)

For ions (particles that have lost or gained electrons) the charge needs to be taken into account.

Beryllium atoms $_{4}^{9}\text{Be}$ protons = 4, electrons = 4, neutrons = 9 − 4 = 5

Fluoride ions $_{9}^{19}\text{F}^{-}$ protons = 9, electrons = 9 + 1 = 10, neutrons = 19 − 9 = 10

Calcium ions $_{20}^{40}\text{Ca}^{2+}$ protons = 20, electrons = 20 − 2 = 18, neutrons = 40 − 20 = 20

Isotopes

The chemistry of an atom is determined by how its electrons behave. All the isotopes of an element have the same number of electrons and therefore the same electron arrangement. This means all the isotopes of an element will react chemically in an identical way. However the difference in the number of neutrons between isotopes may cause slight variations in physical properties such as boiling point. Isotopes of the same element will be chemically identical.

There are two isotopes of chlorine

$_{17}^{37}\text{Cl}$ protons = 17, electrons = 17, neutrons = 20

$_{17}^{35}\text{Cl}$ protons = 17, electrons = 17, neutrons = 18

1.3 The arrangement of electrons
Specification reference: 3.1.1

Nature of electrons

Over time, chemists' understanding of electrons has improved, as technology has got better. Chemists today believe that electrons have some properties of particles, some of waves, and at other times behave as clouds of charge.

Energy levels

Electrons are in constant motion around the nucleus of an atom.

- Electrons are found in energy levels or shells.
- These are split into sub-levels or sub-shells with different maximum numbers of electrons.
- Different types of sub-level contain different numbers of orbitals.
- Each orbital can hold two electrons (one spinning up and one spinning down).
- The shape of an orbital tells you where an electron is most likely to be found.

▼ **Table 1** *Deducing the maximum number of electrons in each energy level*

Energy level	1	2		3			4			
Type of sub-level	s	s	p	s	p	d	s	p	d	f
Number of orbitals in sub-level	1	1	3	1	3	5	1	3	5	7
Maximum number of electrons in sub-level	2	2	6	2	6	10	2	6	10	14
Maximum number of electrons in level	2	8		18			32			

Electron configuration

▼ **Table 2** *Deducing spin diagrams*

Element	Electron configuration	Spin diagrams					
		1s	2s	2p	3s	3p	
H	$1s^1$	↑					
He	$1s^2$	↑↓					
Li	$1s^2 2s^1$	↑↓	↑				
C	$1s^2 2s^2 2p^2$	↑↓	↑↓	↑ ↑			
O	$1s^2 2s^2 2p^4$	↑↓	↑↓	↑↓ ↑ ↑			
P	$1s^2 2s^2 2p^6 3s^2 3p^3$	↑↓	↑↓	↑↓ ↑↓ ↑↓	↑↓	↑ ↑ ↑	

- The period that an element is in determines its highest energy level.
- The group that an element is in determines its outer electron configuration.

Synoptic link
This will be useful when you study electron arrangements in Topic 1.5, Electron arrangements and ionisation energy

Key term
Orbital: An orbital is a region where up to two electrons can exist.

Revision tip
The 2p sub-level or sub-shell can contain up to six electrons but a 2p orbital still only contains up to two electrons.

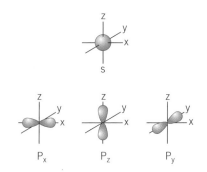

▲ **Figure 1** *Shapes of orbitals*

Summary questions

1 Sketch the shape of an s-orbital. *(1 mark)*

2 Write the full electron configuration for:
   a Na   **b** S   **c** Cl   **d** Be   *(4 marks)*

3 Write full electron configurations for:
   a magnesium, Mg^{2+} ion **b** sulfide, S^{2-} ion
   c oxide, O^{2-} ion
   d sodium, N^+ ion   *(4 marks)*

1.4 The mass spectrometer

Specification reference: 3.1.1

Synoptic link

Mass spectrometry can also be used to measure relative molecular masses and much more, as you will see in Topic 16.2, Mass spectrometry.

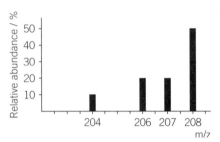

▲ **Figure 1** *The mass spectrum of lead*

Synoptic link

You will find out more about relative atomic mass in Topic 2.1, Relative atomic and molecular masses, the Avogadro constant, and the mole.

Summary questions

1 Name the six steps involved in electro spray ionisation time of flight mass spectrometer.
   (1 mark)

2 Use the spectral data of neon given below to determine its relative atomic mass. Give your answer to an appropriate number of significant figures.

m/z	Relative abundance (%)
20.0	90.9
21.0	0.300
22.0	8.80

   (2 marks)

Mass spectrometer

The mass spectrometer is a very sensitive machine used to analyse samples of elements in terms of the isotopes they contain and their relative amounts and also to discover information about the structure of organic molecules. You must learn names and brief explanations for each step in the electro spray ionisation time of flight (TOF) mass spectrometer.

- **Vacuum** The whole apparatus is kept under vacuum so that no air is present. This stops the ions formed in the apparatus from colliding with molecules in the air.

- **Ionisation** The sample is dissolved in a volatile solvent and then forced through an electrically charged thin hollow needle to produce a stream of positively charged droplets. The solvent dissolves leaving single positively charged ions.

- **Acceleration** The positive ions accelerate towards the negatively charged plate.

- **Ion drift** The positive ions pass through a hole in the negatively charged plate. They form a beam that travels through the flight tube.

- **Detection** The positively charged ions reach the detector and are recorded. Lighter ions travel more quickly so take less time to reach the detector.

- **Data analysis** The information from the detector is sent to a computer which produces a mass spectrum.

Relative atomic mass $\quad A_r = \dfrac{\text{mean mass of an atom of the element}}{\text{mass of one atom of }{}^{12}C} \times 12$

Relative molecular mass $\quad M_r = \dfrac{\text{mean mass of a molecule}}{\text{mass of one atom of }{}^{12}C} \times 12$

Relative isotopic mass $\quad = \dfrac{\text{mass of one atom of the isotope}}{\text{mass of one atom of }{}^{12}C} \times 12$

Relative atomic mass, A_r

The mass of all atoms is measured relative to the mass of a ${}^{12}C$ atom. Relative atomic mass can be calculated using the data from a mass spectrum. From the mass spectrum of lead opposite we obtain the following data:

m/z	204	206	207	208
Relative abundance / %	10.0	20.0	20.0	50.0

- *m/z* is the same as the mass of the ion if the charge is +1.
- Relative abundance tells us the proportion of each isotope present in the sample.
- The A_r of lead can then be calculated as a weighted mean average.

$$A_r = \frac{(204 \times 10.0) + (206 \times 20.0) + (207 \times 20.0) + (208 \times 50.0)}{100} = 207$$

When answering calculation questions it is essential to record your answer to an appropriate number of significant figures. The data in the question is given to three significant figures so the final answer must also be recorded to three significant figures.

1.5 Electron arrangements and ionisation energy

Specification reference: 3.1.1

Ionisation energy

The values obtained for ionisation energies provide substantial evidence for the existence of energy levels and sub-levels.

First ionisation energies of Group 2 elements

Down the group, the nuclear charge (number of protons in the nucleus) increases. However, the electrons in the outer shell are lost more easily. This is because down the group the electrons in the outer shell are further from the nucleus and there is more shielding. As a result the first ionisation energy decreases as you go down a group.

Successive ionisation energies of magnesium

Successive ionisation energies provide further evidence for the existence of energy levels.

*   There is a general increase in the energy needed to remove each electron from magnesium.

*   This is because the electron is being removed from an ion with an increasing positive charge.

*   There is a very big increase between the 2nd and 3rd ionisation energies this is because the 3rd electron is being removed from an electron shell closer to the nucleus.

*   The group to which an element belongs can be determined by identifying where the big jump in ionisation energy occurs.

First ionisation energies of the elements in Period 3

The graph of first ionisation energies for Period 3 elements provides substantial support for the existence of energy sub-levels.

*   The number of protons increases across period 3 so there is an increase in the charge on the nucleus. As a result the force of attraction between the nucleus and the outer electron increases.

*   The number of electrons also increases but these go into the same energy level so are at a similar distance from the nucleus and experience similar shielding. As a result there is an increase in first ionisation energy across Period 3.

Drops in ionisation energy occur at two points on the graph.

*   Between magnesium and aluminium there is a decrease in ionisation energy because the outer electron in aluminium is in a p sub-level which is of slightly higher energy than the s sub-level, so is easier to remove.

    Mg $1s^2\ 2s^2\ 2p^6\ 3s^2$ Al $1s^2\ 2s^2\ 2p^6\ 3s^2\ 3p^1$

*   Between phosphorus and sulfur there is a slight decrease because in sulfur two of the p-electrons are paired and this pair repel each other and the outer electron is easier to remove.

    P $1s^2\ 2s^2\ 2p^6\ 3s^2\ 3p^3$ S $1s^2\ 2s^2\ 2p^6\ 3s^2\ 3p^4$

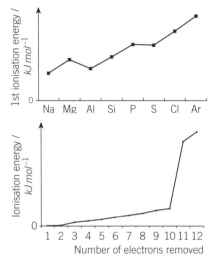

▲ **Figure 1** (a) Successive ionisation energies of magnesium (b) First ionisation energies of the Period 3 elements

Key term

Ionisation energy: First ionisation energy is the energy required to remove 1 electron from each atom in 1 mole of gaseous atoms forming 1 mole of ions with a single positive charge.

e.g. The equation for the first ionisation (of magnesium) is:

$$Mg(g) \rightarrow Mg^+(g) + e^-$$

Summary questions

1   Write down the equation that represents the third ionisation energy of magnesium, Mg.
    (2 marks)

2   Explain why the first ionisation energies of Group 2 elements decrease down the group. *(2 marks)*

3   Sketch a graph showing the successive ionisation energies of oxygen. Label the graph in detail, explaining the patterns it shows. *(2 marks)*

Chapter 1 Practice questions

1. Use the Periodic Table to deduce the full electron configuration of: *(2 marks)*

   a. Mg

   b. Na^+

2. A sample of magnesium was analysed and found to contain three isotopes. The percentage abundance of each isotope is shown below.

Isotope	Percentage abundance / %
^{24}Mg	78.0
^{25}Mg	10.0
^{26}Mg	12.0

   Use the information in the table to calculate the relative atomic mass of this sample of magnesium. Give your answer to three significant figures. *(2 marks)*

3. Complete the table to show the relative charge and relative mass of these sub-atomic particles. *(3 marks)*

Sub-atomic particle	Relative charge	Relative mass
proton		
neutron		
electron		

4. A non-metallic element can be recognised from its relative atomic mass. Analysis of a sample of the non-metallic element revealed the following percentage abundances.

Isotope	% abundance	Relative isotopic mass
1	25.0	37.0
2	75.0	35.0

   a. Define the term isotope. *(1 mark)*

   b. Name the analytic method used to determine the percentage abundance of the non-metallic element. *(1 mark)*

   c. Calculate the relative atomic mass of the non-metallic element. Give your answer to two significant figures. *(2 marks)*

   d. Use your answer and the data sheet to suggest the identity of the non-metallic element. *(1 mark)*

5. Which of these atoms has the smallest atomic radius?

   Ar
   P
   Al
   Na *(1 mark)*

6. Consider the table below. Which line shows the correct number of protons, neutrons, and electrons in the ion?

	Ion	Protons	Neutrons	Electrons
A	$^{23}Na^+$	11	12	11
B	$^{19}F^-$	9	9	8
C	$^{16}O^{2-}$	8	8	8
D	$^{27}Al^{3+}$	13	14	10

   (1 mark)

2.1 Relative atomic and molecular masses, the Avogadro constant, and the mole

Specification reference: 3.1.2

Calculations using relative molecular mass and relative formula mass

a Calculate the relative molecular mass of oxygen, O_2.

$$M_r = 16.0 \times 2 = 32.0$$

b Calculate the relative molecular mass of carbon dioxide, CO_2.

$$M_r = 12.0 + (16.0 \times 2) = 44.0$$

c Calculate the relative formula mass of sodium chloride, NaCl.

$$M_r = 23.0 + 35.5 = 58.5$$

d Calculate the relative formula mass of calcium hydroxide, $Ca(OH)_2$.

$$M_r = 40.1 + (17.0 \times 2) = 74.1$$

The mole and the Avogadro constant

Chemists are interested in how many atoms, molecules, and ions take part in reactions. All of these particles are very small so you cannot determine their mass. Instead you carry out calculations using the concept of the mole.

- The mol is the unit for amount of substance.
- One mole of a substance contains the same number of particles as there are atoms in exactly 12 g of ^{12}C.
- The number of particles in one mole of a substance is the Avogadro number (after Amedeo Avogadro), 6.022×10^{23}.

Amount of substance

The amount of a substance is readily calculated from the relative atomic mass or the relative molecular mass of a substance using the relationship: mass = $M_r \times n$. You must be able to manipulate this relationship in order to calculate mass, M_r, or n given suitable data. n is the number of moles of the substance.

Calculations using mass = $M_r \times n$

a Calculate the mass in grams of 2 moles of Mg atoms.

$$\text{mass} = A_r \times n = 24.3 \times 2 = 48.6\,\text{g}$$

b Calculate the number of moles of Na atoms in 6.5 g of Na atoms.

$$n = \frac{\text{mass}}{A_r} = \frac{6.5}{23.1} = 0.28$$

c Calculate the number of moles of MgO in 2×10^{-6} g of MgO.

$$n = \frac{\text{mass}}{M_r} = \frac{2 \times 10^{-6}}{40.3} = 4.96 \times 10^{-8}$$

d Calculate the M_r of aluminium oxide, Al_2O_3, if 0.5 moles of aluminium oxide has a mass of 51 g

$$M_r = \frac{\text{mass}}{n} = \frac{51}{0.5} = 102$$

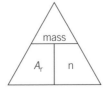

A_r = relative atomic mass

n = number of moles

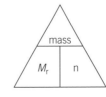

M_r = relative molecular mass / formula mass

n = number of moles

▲ **Figure 1** *Amount of substance*

Key term

Relative atomic mass: The relative atomic mass, A_r, is the weighted average mass of an atom of an element, taking into account its naturally occurring isotopes, relative to $\frac{1}{12}$ of the relative atomic mass of an atom of carbon-12. The relative atomic mass of an element is displayed on the Period Table.

Key term

Relative molecular mass: The relative molecular mass, M_r, of a molecule is the mass of that molecule compared to $\frac{1}{12}$ of the relative atomic mass of an atom of carbon-12. The relative molecular mass of a substance is found by adding together the relative atomic mass of all the atoms in the molecule.

Key term

Relative formula mass: The term relative formula mass is used for ionic compounds as ionic compounds do not contain molecules. Relative formula mass also has the symbol M_r. It is calculated using the same method used for relative molecular mass. Relative atomic mass, relative molecular mass, and relative formula mass do not have units.

Summary questions

1 Calculate the following. Remember to show all your working. Give your answers to an appropriate number of significant figures.

a The number of moles of CaO in 5.61 g of CaO

b The mass of 0.150 moles of K

c The mass of 0.32 moles of LiOH *(3 marks)*

2.2 Moles in solution

Specification reference: 3.1.2

Revision tip

Make sure you give your answer to the required number of decimal places.

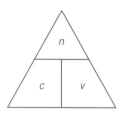

n = number of moles
c = concentration in mol dm^{-3}
v = volume of solution in dm^3

▲ **Figure 1** *Concentration is the amount of a substance in moles dissolved in 1 dm^3 of solution*

Concentrations

- Concentration is the amount of a substance in moles dissolved in 1 dm^3 of solution (1 dm^3 = 1000 cm^3). Concentration has units of mol dm^{-3}.
- This means that a 2.00 mol dm^{-3} solution of sulfuric acid contains 2.00 moles of sulfuric acid dissolved in 1.00 dm^3 of water. As the M_r of sulfuric acid is 98.1.
- This is 2.00 × 98.1 = 196.2 g of sulfuric acid dissolved in 1 dm^{-3} of solution.

The relationship between concentration and number of moles is: $n = c \times v$
n = number of moles, c = concentration in dm^3, v = volume of solution in dm^3
Note – The volume used in the concentration expression is needed in dm^3. To convert dm^3 to cm^3 you must multiply by 1000. To convert cm^3 to dm^3 you must divide by 1000.

Concentration calculations

a Calculate the number of moles of hydrochloric acid in 50 cm^3 of 0.1 mol dm^{-3} solution.

Number of moles = $\frac{50}{1000} \times 0.1 = 0.005$ mol

b Calculate the volume, in cm^3, of 0.0500 mol dm^{-3} sodium hydroxide solution that would contain 1.15 × 10^{-3} moles.

Volume = $1.15 \times 10^{-3} \times \frac{1000}{0.0500} = 23.0$ cm^3

c 50.0 cm^3 of nitric acid contains 1.25 × 10^{-3} moles. Calculate the concentration of the acid.

Concentration = $1.25 \times 10^{-3} \times \frac{1000}{50.0} = 0.250$ mol dm^{-3}

Acid–base titration calculations

Follow these systematic steps to carry out a titration calculation.

25.0 cm^3 of sodium hydroxide solution is exactly neutralised by 21.4 cm^3 of 0.500 mol dm^{-3} hydrochloric acid. What was the concentration of the sodium hydroxide solution?

Step 1	Write a balanced equation NaOH(aq) + HCl(aq) → NaCl(aq) + H$_2$O(l)
Step 2	Write out the data given in the question under the equation. NaOH(aq) + HCl(aq) → NaCl(aq) + H$_2$O(l) 25.0 cm^3 21.4 cm^3 ? 0.500 mol dm^{-3}
Step 3	Convert the substance you know the most about into moles. number of moles of HCl(aq): $n = c \times v = 0.500 \times (21.4 \div 1000) = 0.0107$
Step 4	Use the balanced equation to determine the moles of the unknown substance. 1 mol HCl reacts with 1 mol NaOH. So number of moles of NaOH = 0.0107
Step 5	Convert the number into the units asked for in the question. concentration of sodium hydroxide solution: $c = \frac{n}{v} = \frac{0.0107}{(25.0 \div 1000)}$ = 0.428 mol dm^{-3}

Summary questions

1 1.00 moles of a nitric acid, HNO$_3$, is dissolved in 500 cm^3 of water. What is the concentration of the nitric acid? *(1 mark)*

2 25.0 cm^3 of NaOH(aq) is exactly neutralised by 27.6 cm^3 of 0.150 mol dm^{-3} HNO$_3$(aq). Calculate the concentration of the NaOH(aq) solution. Give your answer to an appropriate number of significant figures. *(3 marks)*

2.3 The ideal gas equation
Specification reference: 3.1.2

The ideal gas equation
Finding the pressure of a gas

What is the pressure exerted by 0.200 moles of chlorine gas in a vessel of 6.00 m³ at a temperature of 298 K?

$$p = \frac{nRT}{V} = \frac{0.200 \times 8.31 \times 298}{6.00} = 82.5 \text{ Pa}$$

Finding the volume of a gas

What is the volume of 6.40×10^{-4} moles of hydrogen gas at a pressure of 1.00 Pa at 273 K?

$$V = \frac{nRT}{P} = \frac{6.40 \times 10^{-4} \times 8.31 \times 273}{1.00} = 1.45 \text{ m}^3$$

Calculating the volume of gas released in a reaction

2.00 g of calcium carbonate is reacted with an excess of 0.200 mol dm⁻³ hydrochloric acid. Calculate the volume of carbon dioxide (in m³) released at standard temperature and pressure.

$$CaCO_3(s) + 2HCl(aq) \rightarrow CaCl_2(aq) + H_2O(l) + CO_2(g)$$

Step 1 Calculate the number of moles of calcium carbonate.

$$n = \frac{\text{mass}}{M_r} = \frac{2.00}{(40.1 + 12.0 + (3 \times 16.0))} = \frac{2.00}{100.1} = 0.01998$$

Step 2 Use the balanced equation to determine the number of moles of carbon dioxide.

$$\text{ratio } CaCO_3 : CO_2 \qquad 1:1$$

$$n \, CO_2 = 0.01998$$

Step 3 Use the ideal gas equation to calculate the volume of carbon dioxide.

$$V = \frac{nRT}{p} = \frac{0.01998 \times 8.31 \times 298}{1.00 \times 10^5} = 4.95 \times 10^{-4} \text{ m}^3$$

Summary questions

1. Nitrogen and hydrogen react in the Haber process to form ammonia. Calculate the volume of ammonia that will be formed when 90.0 cm³ of hydrogen reacts completely with nitrogen if the temperature and pressure are kept constant. $N_2(g) + 3H_2(g) \rightarrow 2NH_3(g)$ *(1 mark)*

2. Calculate the volume occupied by 2.31 moles of hydrogen gas at a pressure of 2×10^5 Pa and a temperature of 32 °C. *(1 mark)*

3. Magnesium metal reacts with hydrochloric acid to form hydrogen gas.
$$Mg(s) + 2HCl(aq) \rightarrow MgCl_2(aq) + H_2(g)$$
Calculate the volume of gas released when 0.20 g of magnesium reacts with an excess of acid at 298 K and 1×10^5 Pa. *(3 marks)*

2.4 Empirical and molecular formula
2.5 Balanced equations and related calculations

Specification reference: 3.1.2

Element	Ca	C	O
% by mass	40.06	11.99	47.95
$\div A_r$	0.999	0.999	2.997
$\div$ smallest	1	1	3

Calculating empirical formula

A compound contains 40.06% calcium, 11.99% carbon, and 47.95% oxygen by mass. Determine its empirical formula. Show all your working.

Empirical formula: $CaCO_3$

Calculating the molecular formula

A compound has an empirical formula of CH_2 and a relative molecular mass of 98.0. Calculate the molecular formula of the compound.

The relative mass of the empirical formula = $12.0 + (2 \times 1.0) = 14.0$

$$\frac{\text{the relative formula mass}}{\text{relative mass of the empirical formula}} = \frac{98.0}{14.0} = 7$$

The molecular formula = C_7H_{14}

Writing a balanced equation

Magnesium reacts with oxygen to form magnesium oxide.

Step 1 Work out the identities of the reactants and products.

$$\text{magnesium} + \text{oxygen} \rightarrow \text{magnesium oxide}$$

Step 2 Construct formulae for each of the reactants and products.

$$Mg + O_2 \rightarrow MgO$$

Step 3 Balance the equation so that the number of each type of atom is the same on each side of the equation. Work from left to right.

$$\mathbf{2}Mg + O_2 \rightarrow \mathbf{2}MgO$$

Step 4 Add state symbols using (s) for solid, (l) for liquid, (g) for gas and (aq) for aqueous, a solution in water.

$$2Mg(s) + O_2(g) \rightarrow 2MgO(s)$$

Summary questions

1 Write the balanced symbol equation for the reaction between sodium and chlorine. Include state symbols. *(2 marks)*

2 Calculate the empirical formula of an oxide of iron which contains 69.9% iron by mass. *(3 marks)*

3 A compound is analysed and found to have an empirical formula of CH_2 and a relative molecular mass of 42.0. Deduce the molecular formula of the compound. *(2 marks)*

2.6 Balanced equations, atom economies, and percentage yields

Specification reference: 3.1.2

Balanced chemical equations in unfamiliar situations

You will be expected to write and balance simple equations for reactions that you have studied in the unit. You may also be given an unfamiliar equation and be asked to balance it.

Balancing an equation for an unfamiliar reaction

Ammonia, NH_3, reacts with sodium to form sodium amide, $NaNH_2$, and hydrogen. Write a balanced equation for this reaction.

Step 1 Write formulae for the reactants on the left of an equation and for the products on the right.

$$NH_3 + Na \rightarrow NaNH_2 + H_2$$

Step 2 Work through the equation from left to right balancing each atom in turn. There is one nitrogen atom on the left and one on the right so this equation balances in terms of nitrogen.

There are three hydrogen atoms on the left and four on the right. In order to provide enough hydrogen atoms for the right you must place a two in front of the ammonia.

$$\mathbf{2}NH_3 + Na \rightarrow NaNH_2 + H_2$$

You now need to check the nitrogen atoms again; there are two on the left so you need to place a two in front of the sodium amide.

$$\mathbf{2}NH_3 + Na \rightarrow \mathbf{2}NaNH_2 + H_2$$

The nitrogen and hydrogen atoms are now balanced. To finish the equation you need to place a two in front of the sodium atom.

$$\mathbf{2}NH_3 + \mathbf{2}Na \rightarrow \mathbf{2}NaNH_2 + H_2$$

Atom economy

The atom economy of a chemical reaction is the proportion of reactants that are converted into useful products.

- Processes with high atom economy are more efficient and produce less waste.
- This is important for sustainable development.

Atom economy is calculated by looking at the mass of desired product in relation to the total mass of the products. These may be expressed as mass in grams or as relative molecular mass. The total mass of the reactants is equal to the total mass of the products.

Calculating atom economy

a A chemical reaction produces 48 g of a desired product and 27 g of undesired product. Calculate the percentage atom economy.

$$\% \text{ atom economy} = \frac{\text{mass of desired product}}{\text{total mass of all products}} \times 100$$

$$= \frac{48}{75} \times 100 = 64\%$$

> ### Key term
>
> **Atom economy:**
>
> $\% \text{ atom economy} =$
> $$\frac{\text{mass of desired product}}{\text{total mass of all products}} \times 100$$

b The reaction of methane with steam produces hydrogen gas along with carbon monoxide as a waste product. Calculate the percentage atom economy in relation to hydrogen gas.

$$CH_4(g) + H_2O(g) \rightarrow 3H_2(g) + CO(g)$$

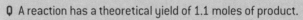

$$\% \text{ atom economy} = \frac{\text{relative mass of desired product}}{\text{total relative mass of all products}} \times 100$$

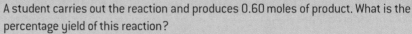

$$= \frac{3 \times 2}{((3 \times 2) + 28)} \times 100 = \frac{6}{34} \times 100 = 17.6\%$$

Note that the ratios of each reacting species have been taken into account here.

Percentage yield

The actual yield of a product shown as a percentage of the expected yield.

$$\text{percentage yield} = \frac{\text{actual yield}}{\text{theoretical yield}} \times 100$$

Worked example

Q A reaction has a theoretical yield of 1.1 moles of product.

A student carries out the reaction and produces 0.60 moles of product. What is the percentage yield of this reaction?

Q Percentage yield $= \frac{0.60}{1.1} \times 100 = 55\%$

Summary questions

1 Ethene, C_2H_4, reacts with hydrogen bromide, HBr, to form bromoethane, C_2H_5Br. Calculate the percentage yield for this reaction if 2.00 g of ethene forms 5.80 g of bromoethane. *(3 marks)*

2 Chlorine gas can be obtained from the electrolysis of brine. The equation for this process is:
$2NaCl(aq) + 2H_2O(l) \rightarrow 2NaOH(aq) + Cl_2(g) + H_2(g)$.
Calculate the atom economy for producing chlorine. *(2 marks)*

3 The Haber process for making ammonia by reacting nitrogen and hydrogen gases
$N_2(g) + 3H_2(g) \rightleftharpoons 2NH_3(g)$
typically has a percentage yield of around 15%. Explain what is meant by the term percentage yield and compare this with the atom economy for this process. *(2 marks)*

Chapter 2 Practice questions

1   A 10.0 g sample of metal was analysed and found to contain 20.0% iron by mass.

   a Calculate the amount, in mol, of iron in the sample. Give your answer to an appropriate number of significant figures.   *(1 mark)*

   b Calculate the number of atoms of iron in the sample of metal.

      $N_A = 6.02 \times 10^{23}\,\text{mol}^{-1}$ *(1 mark)*

2   A chloride of phosphorus was analysed and found to contain 14.9% phosphorus by mass.

   a Define the term *empirical formula*.   *(1 mark)*

   b Calculate the empirical formula of the compound.   *(2 marks)*

3   25.0 cm³ of a solution of sodium hydroxide of concentration $0.100\,\text{mol dm}^{-3}$ was reacted with a dilute solution of hydrochloric acid of unknown concentration.

   23.5 cm³ of the hydrochloric acid was required for complete neutralisation.

   The equation for the reaction is shown below.

   $HCl(aq) + NaOH(aq) \rightarrow NaOH(aq) + H_2O(l)$

   a Calculate the amount, in mol, of sodium hydroxide in the 25.0 cm³ sample.   *(2 marks)*

   b Calculate the amount, in mol, of hydrochloric acid required.   *(1 mark)*

   c Calculate the concentration of the hydrochloric acid solution.   *(2 marks)*

4   A student placed 1.50 g of zinc metal into a beaker containing 50 cm³ of $0.2\,\text{mol dm}^{-3}$ hydrochloric acid.

   The equation for the reaction is shown below.

   $Zn(s) + 2HCl(aq) \rightarrow ZnCl_2(aq) + H_2(g)$

   a What would you see during the reaction?   *(2 marks)*

   b Calculate the amount of zinc added.   *(2 marks)*

   c Calculate the amount of hydrochloric acid in the beaker.   *(2 marks)*

   d Would there be any zinc left at the end of the experiment? Explain your answer.   *(1 mark)*

5   How many atoms are present in 5.00 g of calcium carbonate, $CaCO_3$?

   $N_A = 6.02 \times 10^{23}\,\text{mol}^{-1}$ *(1 mark)*

   A 3.01×10^{22}

   B 1.50×10^{23}

   C 0.0500

   D 9.03×10^{22}

6   What is the amount, in mol, in 50.0 cm³ of a $0.200\,\text{mol dm}^{-3}$ solution of copper(II) sulfate?   *(1 mark)*

   A 10 mol

   B 0.0100 mol

   C 0.0250 mol

   D 0.002 mol

7   Ethanol, C_2H_5OH, can be produced by reacting ethene, C_2H_4, with steam, H_2O. The equation for the reaction is shown below.

   $C_2H_4(g) + H_2O(g) \rightarrow C_2H_5OH(g)$

   Deduce the atom economy of this reaction.

   A 50%   **B** 100%   **C** 46%   **D** 61%   *(1 mark)*

3.1 The nature of ionic bonding

Formation of ions

All chemical bonds are forces of attraction. Ionic bonds occur when a metal and a non-metal react to form a compound.

- Metal atoms lose electrons forming cations. For example, a lithium atom loses 1 electron to form a lithium ion.

  $Li \rightarrow Li^+ + e^-$ The lithium ion has electron configuration $1s^2$.

- Non-metal atoms gain electrons forming anions. For example, a fluorine atom gains 1 electron to form a fluoride ion.

  $F + e^- \rightarrow F^-$ The fluoride ion has electron configuration $1s^2\ 2s^2\ 2p^6$.

Ionic lattices

Ionic compounds are made up of lattice structures.

- In an ionic compound each ion is surrounded by ions of opposite charge.
- For example, in the sodium chloride lattice, each chloride ion is surrounded by six sodium ions and vice versa.
- This structure is repeated throughout the ionic compound.
- As a result ionic compounds are said to have giant structures.

Constructing ionic formulae

Ionic formulae show the ratio of cations to anions present in the ionic lattice. A formula for an ionic compound is constructed from the formulae of the cations and anions present. In most cases you can work out the charges of the cation and anion using the Periodic Table.

- For metal atoms the charge on the ion is the same as the group number of the metal.
- Ionic compounds have a neutral charge overall so the number of positive charges must match the number of negative charges. For example: Barium chloride contains Ba^{2+} ions and Cl^- ions, so has the formula $BaCl_2$.
- Some ionic compounds contain compound ions or molecular ions.

For example:

- Sodium sulfate contains Na^+ ions and SO_4^{2-} ions, so has the formula Na_2SO_4. Aluminium sulfate contains Al^{3+} ions and SO_4^{2-} ions, so has the formula $Al_2(SO_4)_3$.

Properties of ionic compounds ⚗

- Ionic compounds have giant ionic structures. They always have high melting points and are solid at room temperature. Ionic compounds conduct electricity when molten or dissolved as the ions can move. However, they do not conduct electricity when solid as the ions cannot move. Ionic compounds are brittle and shatter easily when hit.

3.2 Covalent bonding

Specification reference: 3.1.3

Sharing of electrons

Non-metal atoms can achieve full outer shells either by accepting electrons from metal atoms or by sharing pairs of electrons.

- Covalent bonds can form between identical atoms or different atoms.
- As with ionic bonding only the outer energy level electrons are involved in bonding.
- A pair of electrons is shared between two atoms.
- The bond is held together by the attraction between each nucleus involved in the bond and the pair of electrons.

Covalent bonds can be represented by dot-and-cross diagrams. They can also be shown as a straight line between the atoms. This is called a displayed formula.

Multiple bonds

Non-metal atoms can form double or triple covalent bonds by sharing more than one pair of electrons. These are represented in molecular formulae by multiple lines between the atoms. Double and triple bonds are stronger than single bonds.

Bonding with lone pairs

Lone pairs of electrons are able to form dative covalent bonds with atoms that have vacant orbitals. Dative covalent bonds are shown in displayed formulae by an arrow.

For example:

- The ammonium ion, NH_4^+, has a dative covalent bond between the nitrogen atom and one of the hydrogen atoms.
- In carbon monoxide, CO, the oxygen atom forms a double covalent bond and also a co-ordinate bond.
- In the NH_3BH_3 molecule the boron atom has a vacant orbital and accepts a pair of electrons from the nitrogen atom.

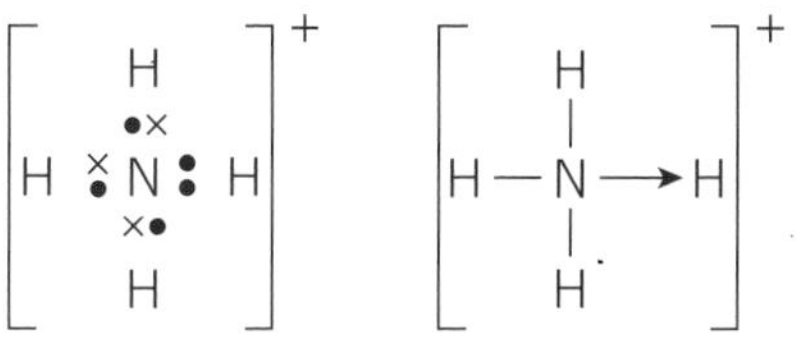

▲ **Figure 1** *Examples of molecules containing dative covalent bonds*

Simple molecular substances 🧪

Simple molecular substances can be solid, liquid, or gas at room temperature but they tend to have relatively low melting points. There are strong bonds within the molecules but only weak forces of attraction between molecules. Simple molecules are poor electrical conductors. Some simple molecules dissolve in water.

Key terms

Single covalent bond: A shared pair of electrons.

Double covalent bond: Two shared pairs of electrons.

Triple covalent bond: Three shared pairs of electrons.

Key terms

Lone pair of electrons: A pair of electrons not involved in bonding.

Dative covalent (or co-ordinate) bond: A shared pair of electrons in which one of the atoms contributes both electrons.

Summary questions

1. What is a covalent bond? *(2 marks)*

2. What is a dative covalent bond? *(2 marks)*

3. Why is nitrogen, N_2, a gas at room temperature? *(2 marks)*

3.3 Metallic bonding

Specification reference: 3.1.3

The metallic bond

Atoms of metals are held together by metallic bonds. In a metallic bond:

- Each metal atom forms a positive ion (cation).
- The positive ions are arranged into a regular lattice structure.
- The ions in the structure are very close to each other so the electrons that are lost when the metal atoms form ions become delocalised.
- Delocalised electrons are not attracted to any particular ion.
  - As a result, these electrons are free to move through the metal.

For example, magnesium atoms have the electron configuration $1s^2\ 2s^2\ 2p^6\ 3s^2$. The 3s electrons are delocalised from each atom forming magnesium ions of charge 2+. These have the electron configuration $1s^2\ 2s^2\ 2p^6$.

The metallic bond is the strong attraction between the positive ions in the lattice and the delocalised sea of electrons.

Metallic bond strength

Metallic bonds do not all have the same strength. If they did, then all metals would melt and boil at the same temperature.

The strength of a metallic bond is dependent on:

- the charge of the ions in the lattice
- the number of electrons in the sea of delocalised electrons.

The boiling point graph for sodium, magnesium, and aluminium shows this. Magnesium needs more energy to boil than sodium as the attraction between the Mg^{2+} ions in the lattice and the sea of electrons is greater than between the Na^+ and the sea of electrons. The force of attraction is even stronger in aluminium, which has Al^{3+} ions.

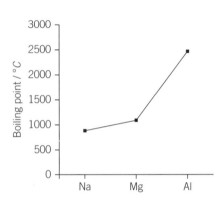

▲ **Figure 1** *Boiling points of sodium, magnesium, and aluminium*

Properties of metals

- Metals are good conductors of heat and electricity.
- Metals have high melting points and boiling points.
- Metals are malleable and ductile as the layers of ions can slip over each other.

pulling force

▲ **Figure 2** *Metals are malleable and ductile*

Summary questions

1 Draw a simple sketch of the structure of a piece of aluminium. Label your sketch to show how aluminium is able to conduct electricity.
(2 marks)

2 List the following metals in order of increasing melting point. Explain the order you have chosen: sodium, aluminium, magnesium. *(2 marks)*

3 Consider the term *ductile*. Explain why metals are ductile. *(1 mark)*

3.4 Electronegativity – bond polarity in covalent bonds
Specification reference: 3.1.3

Charge distribution in covalent molecules

The covalent bond is held together by the attraction between the nuclei of the two atoms involved in the bond and the pairs of electrons.

- If the bond is between identical atoms this sharing is equal, e.g. H–H, Cl–Cl
- If the atoms are different then the sharing may be unequal, e.g. H–F. The fluorine atom is much better at attracting the pair of electrons than hydrogen.

Electronegativity

If the two atoms in a bond have different electronegativites then the more electronegative element has a greater share of the electrons. Non-metal elements at the top of Groups 5, 6, and 7 such as N, O, F, and Cl are the most electronegative.

Polar bonds

Polar bonds are those in which the pair of electrons is not shared equally because there is a large difference in electronegativity between the atoms.

- The more electronegative element has a partial negative charge, shown by δ–.
- The less electronegative element has a partial positive charge, shown by δ+.

In the hydrogen chloride molecule:

- The chlorine atom is more electronegative so has a partial negative charge.
- The hydrogen atom therefore has a partial positive charge.
  - As a result, The H–Cl *molecule* can be described as polar.

In the carbon dioxide molecule:

- The oxygen atoms are more electronegative so have a partial negative charge.
- The carbon atom has two partial positive charges as it is bonded to two oxygen atoms. As a result, The C=O *bond* can be described as polar.

H 2.2							He –
Li 1.0	Be 1.6	B 2.0	C 2.5	N 3.0	O 3.4	F 4.0	Ne –
Na 0.9	Mg 1.3	Al 1.6	Si 1.9	P 2.2	S 2.6	Cl 3.2	Ar –
K 0.8	Ca 1.0					Br 3.0	Kr 3.0
Rb 0.8						I 2.7	Xe 2.6

▲ Figure 3

Key term

Covalent bonding: Covalent bonding is the sharing of one or more pairs of electrons.

Key term

Electronegativity: Electronegativity is the ability of an atom to withdraw electron density from a covalent bond.

$$\overset{\delta^+}{H} - \overset{\delta^-}{Cl}$$

▲ **Figure 1** *Hydrogen chloride*

$$\overset{\delta^-}{O} = \overset{2\delta^+}{C} = \overset{\delta^-}{O}$$

▲ **Figure 2** *Carbon dioxide*

Key terms

Polar bond: A covalent bond between atoms with different electronegativities.

Dipole: Opposite charges separated by a short distance in a molecule or ion.

Polar bonds and polar molecules

A molecule that contains polar bonds may not be a polar molecule. To find out if a molecule is polar:

- Draw the molecule (in three dimensions if necessary, remembering about the influence of lone pairs).
- Label any polar bonds using the $\delta+$, $\delta-$ convention.
- Then examine the shape of the molecule.
- If the molecule is symmetrical the polar bonds cancel out and the molecule is not polar.
- If the molecule is not symmetrical polar bonds do not cancel out and the molecule is polar.

Worked example

Q. Are carbon dioxide and water polar molecules? Explain your answers.

A. $\overset{\delta-}{O} = \overset{\delta+}{C} = \overset{\delta-}{O}$ Carbon dioxide has two polar bonds but there is no net dipole as the bonds are symmetrical. Carbon dioxide is a non-polar molecule.

    Water has two polar bonds and has a net dipole beacause the polar bonds do not cancel each other out. Water is polar molecule. This has a big influence on the properties of water.

Summary questions

1   Label the following bonds to indicate their polarity.
    a H–F   **b** C–Cl   **c** O–N   **d** S=O   (4 marks)

2   Draw the following molecules. Label any polar bonds and state whether the molecule is polar.
    a hydrogen bromide
    b hydrogen sulfide
    c ammonia
    d fluorine oxide, F_2O (8 marks)

3   Arrange the substances below in order of their covalent character starting with the least covalent:
    a NaCl, $AlCl_3$, $MgCl_2$ **b** NaI, NaCl, NaBr   (2 marks)

3.5 Forces acting between molecules
Specification reference: 3.1.3

Van der Waals' forces

A van der Waals' force is a force of attraction between a temporary dipole on one molecule and an induced dipole on another molecule.

Temporary dipoles

Electrons in a molecule are constantly moving.

- This means that the electron cloud around an atom or within a non-polar molecule is not static.
- At any instant in time the distribution of the electrons may be uneven although on average they are distributed evenly. As a result, a non-polar molecule may have a temporary dipole.

Induced dipoles

The presence of a temporary dipole in one atom or molecule can cause a dipole to form in a nearby atom or molecule. This dipole is called an induced dipole. The induced dipole can then induce a dipole in a neighbouring atom or molecule. The net effect of this is a force of attraction between the particles called a temporary dipole–induced dipole force.

The more electrons a molecule has the stronger the van der Waals' forces between molecules will be and the higher its boiling temperature is.

Permanent dipole–dipole forces

Molecules with a permanent dipole have regions of different electron density within them. These molecules are described as being polar. It is easy to see that if two polar molecules come near each other in space there will be an attraction between them. For the hydrogen chloride molecule in Figure 1, the electronegative chlorine of one HCl molecule will attract the electropositive hydrogen of another.

Hydrogen bonding

Hydrogen bonds are an especially strong permanent dipole–dipole force, which exist between molecules that contain very electronegative elements.

For a hydrogen bond to occur:

- The molecule must have an O, N, or F atom bonded to a hydrogen atom.
- The molecule to which it is attracted must contain an O, N, or F atom.

There are, of course, many other molecules, for example, methanol that are capable of hydrogen bonding.

$$\overset{\delta^+}{H} - \overset{\delta^-}{Cl} \text{-----} \overset{\delta^+}{H} - \overset{\delta^-}{Cl} \text{-----} \overset{\delta^+}{H} - \overset{\delta^-}{Cl}$$

permanent dipole-dipole force

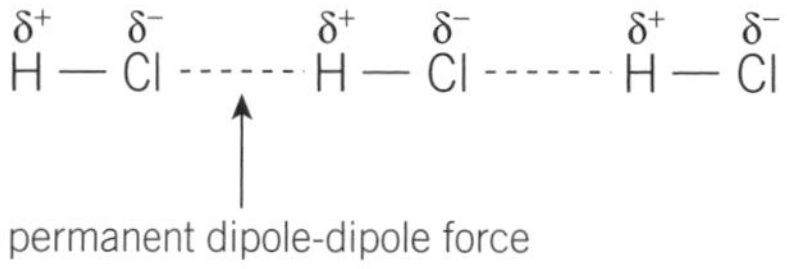

▲ **Figure 1** *Hydrogen chloride*

Summary questions

1   Draw a labelled diagram to show the hydrogen bonding in
    a   hydrogen fluoride
    b   ammonia   (2 marks)

2   Methane, CH_4 and ethane, C_2H_4 are hydrocarbons. Predict which of these hydrocarbons has the highest melting point. Explain your answer.
                                                                   (2 marks)

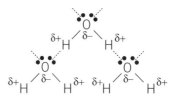

Hydrogen bonding in water

▲ **Figure 2** *Hydrogen bonding in water*

3.6 The shapes of molecules and ions

Specification reference: 3.1.3

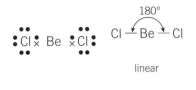

linear

trigonal planar

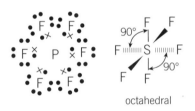

tetrahedral

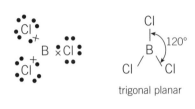

octahedral

▲ **Figure 1** *The shapes of molecules*

Revision tip

The shapes of molecular ions can be worked out in the same way as for uncharged molecules.

- Remember to take account of the charge on the ion when working out the dot-and-cross diagram.
- Positively charged ions have fewer electrons than the original atom.
- Negatively charged ions have gained extra electrons.

Valence shell electron pair repulsion theory

The shapes of molecules and ions are determined by applying valence shell electron pair repulsion theory:

- Valence shell electrons are those in the outermost energy level.
- These pairs of electrons repel each other.
- The shape of a molecule is such that the distance between the pairs of electrons is as large as possible.
- Multiple bonds have the same repelling effect as single bonds.
- Lone pairs of electrons are more repelling than bonding pairs of electrons. As a result the bond angles are slightly smaller when lone pairs are present.

▼ **Table 1** *Predicting the orbital shape and bond angle of molecules*

Pairs of electrons	Basic shape	Bond angles
2	linear	180°
3	trigonal planar	120°
4	tetrahedral	109.5°
6	octahedral	90°

Molecules containing one or more lone pairs of electrons

Tetrahedral

Both ammonia, NH_3, and water, H_2O, are molecules containing four pairs of electrons overall.

- In ammonia one of these pairs of electrons is a lone pair. As a result ammonia has a pyramidal shape with a bond angle of 107°
- In water two of the pairs of electrons are lone pairs. As a result water has a bent (or non-linear) shape with a bond angle of 104.5°.
- The repulsive effect of a lone pair of electrons is greater than a bond pair. As a result the bond angles in ammonia and water are slightly smaller than that of methane.

pyramidal

bent or non-linear

▲ **Figure 2** *Ammonia and water molecules*

Summary questions

1 Draw a dot-and-cross diagram for each of the following molecules. Then use it to draw a labelled diagram of the molecule showing all bond angles. Include the name of the shape of each molecule.
   a BF_3 **b** CCl_4 **c** PH_3 **d** H_2S *(4 marks)*

3.7 Bonding and physical properties
Specification reference: 3.1.3

Changing state 🧪

When a solid is melted or a liquid frozen, a liquid boiled or a gas condensed it is said to have changed state.

- All changes of state involve changes in energy.
- When a solid melts or a liquid boils (or vaporises), energy is used to break the forces between the atoms, molecules, or ions involved.
- As the change of state occurs, the temperature stays constant because the energy provided to the system is used to break the forces of attraction between the particles.

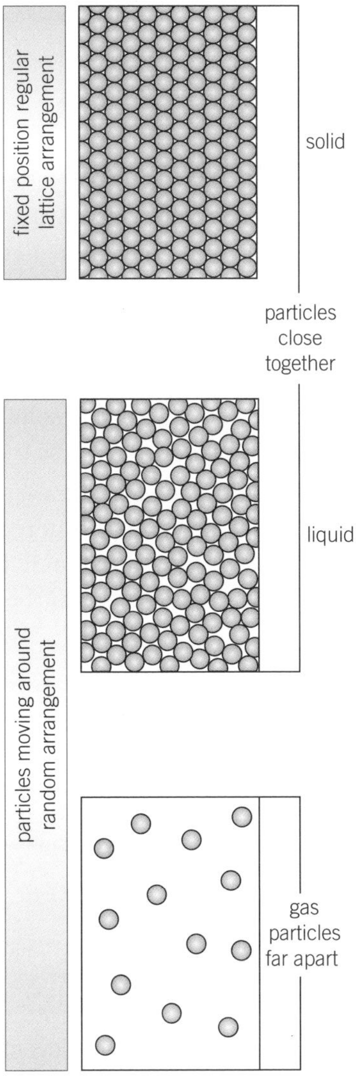

Metals

When melting a pure metal the attraction between the lattice of positive ions and the delocalised sea of electrons is broken. This is a strong force so requires a large amount of energy, which means that metals have high melting points. The melting points increase as the charge on the metal ion and the number of delocalised electrons increases.

Ionic substances

Remember that an ionic bond is an electrostatic attraction between ions of opposite charge. When an ionic substance melts, the energy provided is used to break this attraction. The attraction is strong so ionic substances are hard to melt and are solid at room temperature.

Giant molecular substances

Giant molecular substances have a large network of covalent bonds. Melting these substances involves a large amount of energy as lots of strong covalent bonds must be broken in order to change state. You must be able to explain why diamond and graphite have such high melting points.

Simple molecular substances

Melting simple molecular substances requires the breaking of the intermolecular force between the molecules.

- These forces are much weaker than the covalent bond that exists between the atoms within the molecule.

You must be able to identify the type of force between molecules and relate this to changes of state.

fixed position regular lattice arrangement — solid

particles close together

particles moving around random arrangement — liquid

gas particles far apart

▲ **Figure 1** *States of matter*

Synoptic link
You will learn more about enthalpy in Topic 4.2, Enthalpy.

Revision tip
The energy required to change state from a solid to a liquid is called the enthalpy change of fusion while the energy required to change state from a liquid to a gas is called the enthalpy change of vaporisation.

Revision tip
Simple molecular structures have strong covalent bonds within the molecules but only weak intermolecular forces of attraction between molecules.

▼ **Table 1** *Intermolecular forces*

Name of force	Relative strength of force
permanent dipole–dipole attraction	stronger than van der Waals' force, weaker than hydrogen bond
hydrogen bond	the strongest intermolecular force
van der Waals' force	weakest intermolecular force

Conducting electricity

In order to conduct electricity, charged particles must be able to move through a substance when a voltage is applied:

- In a metal the delocalised electrons are free to move.
- In a molten ionic substance the ions are free to move. Note that ionic solids do not conduct electricity as the ions are held in a fixed position but they do conduct if the solid is dissolved in another substance so that the ions can move.
- Simple molecular substances and giant molecular substances do not conduct electricity as there are no charged particles. The exception to this is carbon – graphite is able to conduct as there are delocalised electrons between the layers of carbon atoms.

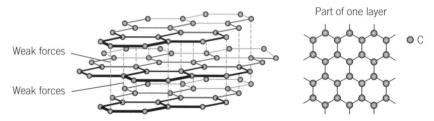

▲ **Figure 2** *The structure of graphite*

Summary questions

1 Complete the table below showing the type of bonding in each substance

Name of substance	Formula	Type of structure	Type of bonding
magnesium	Mg	giant ionic	
sodium chloride			ionic
chlorine			
graphite			

(1 mark)

2 Arrange the following substances in order of melting point starting with the lowest. Explain the order you have chosen.
H_2O, $MgCl_2$, Cl_2.

(2 marks)

Chapter 3 Practice questions

1 The shape of methane, CH_4 and ammonia, NH_3, molecules can be predicted using valence electron pair repulsion theory.

    a Methane, CH_4, has a bond angle of 109.5°. Name the shape of the methane molecule and explain why methane has the bond angle of 109.5°.   *(3 marks)*

    b State and explain the bond and angle in an ammonia, NH_3, molecule.   *(3 marks)*

2 Define the term electronegativity.   *(2 marks)*

3 **a** Describe the bonding and structure of magnesium. Include a diagram in your answer.   *(4 marks)*

    b Explain why magnesium is a good electrical conductor.   *(1 mark)*

4 Water has a higher melting point and boiling point than expected compared with the other Group 6 hydrides. Explain why water has a higher melting point than expected. Include a diagram in your answer.   *(4 marks)*

5 **a** Complete the table below to show the structure, type of bonds, and electrical conductivity of magnesium and magnesium oxide, MgO.   *(3 marks)*

Substance	Mg	MgO
Structure		
Type of bonding		
Electrical conductivity when solid		

    b Explain why magnesium oxide conducts electricity when it is dissolved in water?   *(1 mark)*

    c Explain why magnesium oxide has a high melting point.   *(2 marks)*

6 Strontium oxide is an ionic compound.

    a Define the term *ionic bond*.   *(2 marks)*

    b Draw a dot-and-cross diagram to show the bonding in strontium oxide. Draw the outer shell of electrons only.   *(1 mark)*

    c Explain why strontium oxide has a high melting point.   *(3 marks)*

7 Which one of these substances contains ionic bonds?

    A H_2O

    B Na_2O

    C CO_2

    D Mg   *(1 mark)*

8 Which of these equations shows the most polar carbon-halogen bond?

    A C–Cl

    B C–F

    C C–I

    D C–Br   *(1 mark)*

Exothermic and endothermic reactions

Energy must be taken in to break existing bonds(an endothermic process), whilst energy is given out when new bonds are formed (an exothermic process). During chemical reactions there is often a difference between the amount of energy taken in to break the existing bonds and the amount of energy given out when the new bonds are formed.

Exothermic reactions

- The substances involved in exothermic reactions get hotter because chemical energy is being changed into thermal (heat) energy.

- During the reaction the chemicals lose energy. The energy that is lost by the chemicals is gained by the surroundings. As a result the enthalpy change (ΔH) for an exothermic reaction is always negative.

Endothermic reactions

Some reactions take in more energy to break existing bonds then they release when new bonds are formed. These reactions are described as endothermic.

- The substances involved in endothermic reactions get colder because thermal energy is taken in from the surroundings.

- The energy that is lost by the surroundings is gained by the chemicals. As a result the enthalpy change (ΔH) for an endothermic reaction is always positive.

Standard enthalpy change of formation

- The standard enthalpy of formation, $\Delta_f H^\ominus$, is the enthalpy change when one mole of a compound is formed from its elements in their standard states, under standard conditions.

  As a result, the standard enthalpy change of formation of an element in its standard state is zero.

Standard enthalpy change of combustion

The standard enthalpy of combustion, $\Delta_c H^\ominus$, is the enthalpy change when one mole of a compound is completely burnt in oxygen under standard conditions.

The reaction involving $\Delta_c H^\ominus$ of methane, CH_4, is represented as

$$CH_4(g) + 2O_2(g) \rightarrow CO_2(g) + 2H_2O(g)$$

$$\Delta_c H^\ominus \text{ at } 25°C = -890.3 \text{ kJ mol}^{-1}$$

Summary questions

1   What is $\Delta_f H^\ominus$ of $N_2(g)$?
    (1 mark)

2   What is the equation that represents the standard enthalpy change of combustion of ethanol? Include state symbols.
    (2 marks)

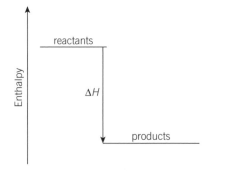

▲ **Figure 1** *An exothermic reaction*

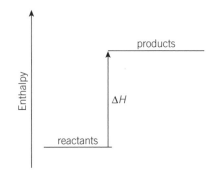

▲ **Figure 2** *An endothermic reaction*

Calorimetry

Worked example

Q A student added an excess of magnesium powder to a solution of $0.100 \, mol \, dm^{-3}$ hydrochloric acid. The student used $50.0 \, cm^3$ of hydrochloric acid. The temperature at the start of the experiment was $21.0 \, °K$. The student measured a maximum temperature of $31.0 \, °K$. Assume that the specific heat capacity of the solution is $4.2 \, J \, g^{-1} \, °K^{-1}$ and that the solution has a density of $1.00 \, g \, cm^{-3}$ Calculate the heat energy change of this reaction.

A $q = mc \Delta T$

$q = 50.0 \times 4.2 \times (31.0 - 21.0) = 2100 \, J$

Notice how the calculated results can only be given to 2 significant figures as the least accurate measurement (specific heat capacity of the solution) was only given to 2 significant figures.

Calculating the enthalpy change of a reaction

The enthalpy change of a reaction is calculated as the heat energy change per mole and has units of $kJ \, mol^{-1}$.

To calculate the enthalpy change of a reaction:

Calculate the heat energy change involved.

Divide the heat energy change for the reaction by 1000.

Calculate the number of moles.

Then divide the heat energy change by the number of moles involved.

Remember to include the sign for the enthalpy change.

Exothermic reactions release energy to the surrounding and have a negative sign.

Endothermic reactions take in energy from the surroundings and have a positive sign.

Revision tip

Calculating the heat change of a reaction

The heat energy change for a reaction (q, measured in joules) is given by the equation

$q = mc \Delta T$

Where m = the mass of the surroundings (g), c = the specific heat capacity of the surroundings, ($J \, g^{-1} \, K^{-1}$) and ΔT is the temperature change (final temperature − initial temperature), (K).

This gives a value for the heat change of the reaction in J.

To give an answer in kilojoules remember to divide by 1000. Normally it is acceptable to give an answer either as joules or kilojoules so it is essential to include the correct unit with your answer.

Worked example

Q A chemist placed $25.0 \, cm^3$ of $0.200 \, mol \, dm^{-3}$ sodium hydroxide solution into a polystyrene cup then added $25.0 \, cm^3$ of $0.200 \, mol \, dm^{-3}$ hydrochloric acid solution. The initial temperature of both solutions was $22.5 \, °C$. The final temperature of the solutions was $29.5 \, °C$. Assume that the specific heat capacity of the solution is $4.18 \, J \, g^{-1} \, K^{-1}$ and that all the solutions have a density of $1.00 \, g \, cm^{-3}$.

Calculate the enthalpy change of the reaction.

A $q = mc \Delta T$

$= 50 \times 4.18 \times 7.0 = 1463 \, J$

$= 1.463 \, kJ$

The number of moles of both reactants is

$\frac{25.0}{1000} \times 0.200 = 0.00500 \, mol$

The enthalpy change = 1.463 kJ/0.00500 mol

$$= -293 \text{ kJ mol}^{-1}$$

The final answer is given to three significant figures and the minus sign added to show the reaction is exothermic.

Measuring enthalpy changes more accurately

A major source of error with this experiment is heat loss. Polystyrene is a good insulator and helps to prevent heat being lost to the surrounding during exothermic reactions or gained from the surrounding by endothermic reactions. Adding a lid to the polystyrene cup would help prevent heat loss but would make it hard to read the thermometer accurately. However adding more insulation to the polystyrene cup would improve the experimental results.

Adding the reactant chemicals together quickly, using powders (which have a high surface area) and stirring the chemicals all ensure that the chemicals react quickly and allow a maximum temperature change to be recorded before the product chemicals start to cool down to (or warm up to) room temperature.

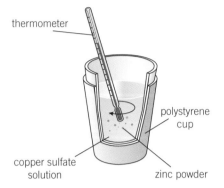

thermometer

polystyrene cup

copper sulfate solution

zinc powder

▲ **Figure 1** *This simple calorimeter can be used to calculate the heat energy change during the displacement reaction between zinc and copper sulfate*

Common misconception

In exothermic reactions heat energy is given out so the temperature of the surroundings increases.

But ΔH of the system is negative.

Summary questions

1 What is the major source of error in a calorimetry experiment carried out in a beaker? How could this error be reduced? (*2 marks*)

2 A student placed 50.0 cm³ of 1.00 mol dm⁻³ copper sulfate solution in a polystyrene cup. The solution has an initial temperature of 21.0 °C. They then quickly added an excess of zinc powder and recorded the maximum temperature as 30.5 °C. Assume that the specific heat capacity of the solution is 4.18 J g⁻¹ K⁻¹ and that all the solutions have a density of 1.00 g cm⁻³
   a Why was zinc powder used?
   b Calculate the heat energy change of this reaction.
   c Calculate the enthalpy change of the reaction. (*5 marks*)

4.4 Hess's law

Specification reference: 3.1.4

Hess's law

The enthalpy change of a chemical reaction is independent of the route by which the reaction is achieved and depends only on the initial and final states.

- As a result you can use Hess's law to find enthalpy changes for reactions that cannot be measured directly in the laboratory.

- The enthalpy change from the reactants A to the products B is the same as the sum of the enthalpy changes from A to C and from C to B.

- We can use $\Delta_f H^\ominus$ and $\Delta_c H^\ominus$ values.

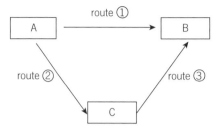

▲ **Figure 1** *route 1 = route 2 + route 3*

Using enthalpies of formation

The standard enthalpy of formation ($\Delta_f H^\ominus$) is the enthalpy change when one mole of a compound is formed from its elements in their standard states under standard conditions.

Standard enthalpies of formation can be used to calculate enthalpy changes that cannot be measured directly.

- Notice that the arrow is drawn from the elements to the chemicals.

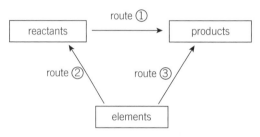

▲ **Figure 2** *Using standard enthalpy of formation values: route 1 = – route 2 + route 3*

Worked example

Q Calculate the enthalpy change for the combustion of ethene, $C_2H_4(g)$.

$$C_2H_4(g) + 3O_2(g) \rightarrow 2CO_2(g) + 2H_2O(g)$$

Given that

$\Delta_f H^\ominus C_2H_4(g) = +52\,kJ\,mol^{-1}$

$\Delta_f H^\ominus CO_2(g) = -394\,kJ\,mol^{-1}$

$\Delta_f H^\ominus H_2O(g) = -242\,kJ\,mol^{-1}$

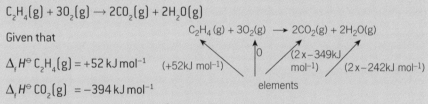

▲ **Figure 3** *The Hess's law diagram for the combustion of ethene*

A The unknown enthalpy change $\Delta_c H^\ominus$

$= - (+52\,kJ\,mol^{-1}) + (2 \times -394\,kJ\,mol^{-1}) + (2 \times -242\,kJ\,mol^{-1})$

$= -1324\,kJ\,mol^{-1}$

Q Calculate the enthalpy change for the reaction below.

$$CH_4(g) + H_2O(g) \rightarrow CO(g) + 3H_2(g)$$

Given that

$\Delta_f H^\ominus CH_4(g) = -75\,kJ\,mol^{-1}$

$\Delta_f H^\ominus CO(g) = -110\,kJ\,mol^{-1}$

$\Delta_f H^\ominus H_2O(g) = -242\,kJ\,mol^{-1}$

$CH_4(g) + H_2O(g) \rightarrow CO(g) + 3H_2(g)$

$-75\,kJ\,mol^{-1} \quad -242\,kJ\,mol^{-1} \quad -110\,kJ\,mol^{-1} \quad 0$

elements

▲ **Figure 4** *The Hess's law diagram for the reaction*

A The unknown enthalpy change

$= -(-75\,mol^{-1}) - (-242\,kJ\,mol^{-1}) + (-110\,kJ\,mol^{-1})$

$= +207\,kJ\,mol^{-1}$

Summary questions

1 What is Hess's law?
(1 mark)

2 In $\Delta_f H^\ominus$ what are the conditions indicated by the symbol $^\ominus$? *(2 marks)*

3 State the equation, including state symbols, that accompanies the enthalpy change of formation of propene, $C_3H_6(g)$. *(2 marks)*

4 Calculate the enthalpy change for the decomposition of magnesium nitrate shown below.
$$Mg(NO_3)_2(s) \rightarrow MgO(s) + 2NO_2(g) + \tfrac{1}{2}O_2(g)$$
Given the information shown in the table:

Substance	$\Delta H_f^\ominus$ $kJ\,mol^{-1}$
$Mg(NO_3)_2(s)$	−791
$MgO(s)$	−602
$2NO_2(g)$	−33

(2 marks)

4.5 Enthalpy changes of combustion

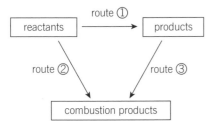

▲ **Figure 1** *Using enthalpy of combustion values: route 1 = route 2 – route 3*

Key term

The standard enthalpy of combustion, $[\Delta_c H^\ominus]$: The enthalpy change when one mole of a compound is completely burnt in oxygen under standard conditions.

Summary questions

1 Define the term *standard enthalpy change of combustion*. **(1 mark)**

2 Write the equation that represents the standard enthalpy change of combustion of propane, $C_3H_8(g)$. **(1 mark)**

3 Use the information in the table to calculate the standard enthalpy change of formation of ethane. **(3 marks)**

$$2C(s) + 2H_2(g) \rightarrow C_2H_4(g)$$

Substance	$\Delta_c H^\ominus kJ\,mol^{-1}$
C(s)	−394
$H_2(g)$	−286
$C_2H_4(g)$	−1409

Using enthalpies of combustion

Hess's law cycles can be used to calculate enthalpy changes that cannot be measured directly. This could happen because the chemicals do not react directly or when a reaction produces several different products. When values for the enthalpy changes of combustion are given in questions the arrow is drawn from the chemicals to the combustion products.

Worked example

Q Calculate the enthalpy change for

$$C(s) + 2H_2(g) \rightarrow CH_4(g)$$

given that

$\Delta_c H^\ominus C(s) = -394\,kJ\,mol^{-1}$

$\Delta_c H^\ominus H_2(g) = -286\,kJ\,mol^{-1}$

$\Delta_c H^\ominus CH_4(g) = -890\,kJ\,mol^{-1}$

A The unknown enthalpy change $\Delta H^\ominus$

$= -394\,kJ\,mol^{-1} + (2 \times -286\,kJ\,mol^{-1}) - (-890\,kJ\,mol^{-1})$

$= -76\,kJ\,mol^{-1}$

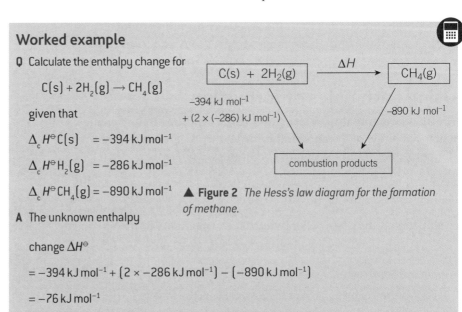

▲ **Figure 2** *The Hess's law diagram for the formation of methane.*

Notice how there are two moles of hydrogen in the equation so the value for $\Delta_c H^\ominus$ must be multiplied by two.

Q Calculate the enthalpy change for the hydrogenation of ethene to form ethane

$$C_2H_4(g) + H_2(g) \rightarrow C_2H_6(g)$$

given that

$\Delta_c H^\ominus C_2H_4(g) = -1409\,kJ\,mol^{-1}$

$\Delta_c H^\ominus H_2(g) = -286\,kJ\,mol^{-1}$

$\Delta_c H^\ominus C_2H_6(g) = -1560\,kJ\,mol^{-1}$

A The unknown enthalpy change $\Delta H^\ominus$

$= -1409\,kJ\,mol^{-1} -286\,kJ\,mol^{-1} - (-1560\,kJ\,mol^{-1})$

$= -135\,kJ\,mol^{-1}$

▲ **Figure 3** *The Hess's law diagram for the hydrogenation of ethene.*

Notice that the method chosen only depends on the values given. Here enthalpy changes of combustion are given so the combustion products are placed at the bottom of the Hess's law diagram and the arrows are drawn downwards.

4.6 Representing thermochemical cycles
Specification reference: 3.1.4

Energetic stability

Enthalpy level diagrams can be used to represent the enthalpy changes of chemical reactions. This enthalpy level diagram (sometimes called an energy level diagram) shows that the products of the reaction, $CO_2(g)$ and $2H_2O(g)$, have a lower energy than the reactants, $CH_4(g)$ and $2O_2(g)$. The products are energetically more stable than the products.

Enthalpy values

Enthalpy changes, ΔH, can be worked out from Hess's law cycles. However, to draw enthalpy level diagrams absolute values of the enthalpies of different substances are required.

The enthalpy of elements

The enthalpies of all elements in their standard states (the state that they exist in at 298 K or 25 °C and 100 kPa) are zero.

298 K or 25 °C and 100 kPa are normal temperature and pressure conditions.

Gases such as oxygen normally exist as oxygen molecules, O_2, so the standard state for oxygen is $O_2(g)$.

Thermochemical cycles and enthalpy level diagrams

The enthalpy change of chemical reactions can be shown using thermochemical cycles or as enthalpy level diagrams. Both of these methods are useful ways to tackle problems.

Notice how in thermochemical cycles the arrows show the direction of the change (from the reactants to the products of a chemical reaction) and are drawn diagonally while in enthalpy level diagrams the arrows are drawn vertically.

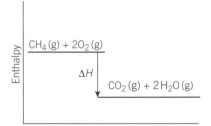

▲ **Figure 1** *An exothermic reaction*

> ### Revision tip
> Some elements exist in several forms in the same state. For example, solid carbon exists as graphite and diamond. These different forms are called allotropes. Under standard conditions the most common form, and therefore the most stable, of the allotropes of carbon is graphite. So graphite is used for the standard state of carbon and is given the symbol C(graphite).

> ### Summary questions
>
> 1 State the enthalpy of nitrogen, $N_2(g)$. *(1 mark)*
>
> 2 Define what is meant by the term *allotrope*. *(1 mark)*
>
> 3 Calculate the standard enthalpy change for the reaction below by using:
>   **a** a thermochemical cycle
>   **b** an enthalpy level diagram
> $2C_2H_6(g) + 7O_2(g) \rightarrow 4CO_2(g) + 6H_2O(g)$
>   You are provided with the following data.
> $4C(s) + 6H_2(g) + 7O_2(g) \rightarrow 4CO_2(g) + 6H_2O(g)$
> $\Delta H^\ominus = -3292$ kJ mol^{-1}
> $4C(s) + 6H_2(g) + 7O_2(g) \rightarrow 2C_2H_6(g) + 7O_2(g)$
> $\Delta H^\ominus = -172$ kJ mol^{-1}
>   *(3 marks)*

Worked example

Q Calculate the standard enthalpy change for the reaction below by using:

  a a thermochemical cycle

  b an enthalpy level diagram

  $2NO(g) \rightarrow N_2O_4(g)$

  You are provided with the following data.

  $N_2(g) + 2O_2(g) \rightarrow 2NO_2(g), \Delta H^\ominus + 33$ kJ mol^{-1}

  $N_2(g) + 2O_2(g) \rightarrow N_2O_4(g), \Delta H^\ominus +9.0$ kJ mol^{-1}

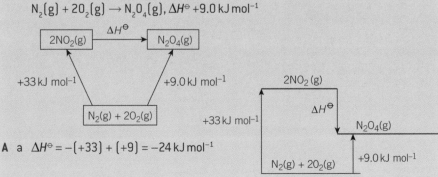

A a $\Delta H^\ominus = -(+33) + (+9) = -24$ kJ mol^{-1}

  b $\Delta H^\ominus = -(+33) + (+9) = -24$ kJ mol^{-1}

Synoptic link

Bond enthalpies give a measure of the strength of bonds, and can help to predict which bond in a molecule is most likely to break. However, this is not the only factor, the polarity of the bond is also important – see Topic 3.5, Forces acting between molecules, and Topic 13.2, Nucleophilic substitution in halogenoalkanes.

Key term

Bond dissociation enthalpies: The enthalpy change required to break one mole of a covalent bond when all the species are in the gaseous state. This is an endothermic reaction.

Revision tip

Bond enthalpies are always endothermic so they always have a positive sign.

Summary questions

1 Define the term *mean bond enthalpy*. (*1 mark*)

2 Give the equation that sums up the reaction for the bond dissociation energy of a H–Cl bond. (*1 mark*)

3 Why does the actual value of bond enthalpy sometimes differ from the mean bond enthalpy for the bond? (*1 mark*)

4 The enthalpy change for the reaction
$CH_4(g) \rightarrow C(g) + 4H(g)$
is +1664 kJ mol^{-1}
What is the average bond enthalpy for a C–H bond? (*1 mark*)

Mean bond enthalpy

The mean (average) bond enthalpy is the mean amount of energy required to break one mole of a specified type of covalent bond in a gaseous species.

Why do values calculated from mean bond enthalpies differ from those calculated from Hess's law?

The mean bond enthalpy for a bond is an average value obtained from many molecules. This means that the actual value for the bond in a particular molecule will probably be a little different from the mean value. So values found using calculations involving mean bond enthalpies will sometimes be slightly different from those obtained from other methods such as from using Hess's law.

Using mean bond enthalpies in calculations

Mean bond enthalpy values can be used to work out the enthalpy change of a reaction.

Worked example

Q. Calculate the enthalpy change when methane, $CH_4(g)$ is burnt.

Bond	Average bond enthalpy / $kJ\,mol^{-1}$
C–H	+413
O=O	+498
C=O	+805
O–H	+464

$CH_4(g) + 2O_2(g) \rightarrow CO_2(g) + 2H_2O(g)$

A Energy required to break the existing bonds

$4 \times$ C–H = 4×413 kJ mol^{-1} = 1652 kJ mol^{-1}

$2 \times$ O=O = 2×498 kJ mol^{-1} = 996 kJ mol^{-1}

Total = 2648 kJ mol^{-1}

Energy released when the new bonds are formed

$2 \times$ C=O = 2×805 kJ mol^{-1} = 1610 kJ mol^{-1}

$4 \times$ O–H = 4×464 kJ mol^{-1} = 1856 kJ mol^{-1}

Total = 3466 kJ mol^{-1}

The net energy change = 2648 kJ mol^{-1} − 3466 kJ mol^{-1} = −818 kJ mol^{-1}

Chapter 4 Practice questions

1 Ethene, C_2H_4 reacts with bromine, Br_2 to form 1,2-dibromoethane.

The equation for the reaction is shown below.

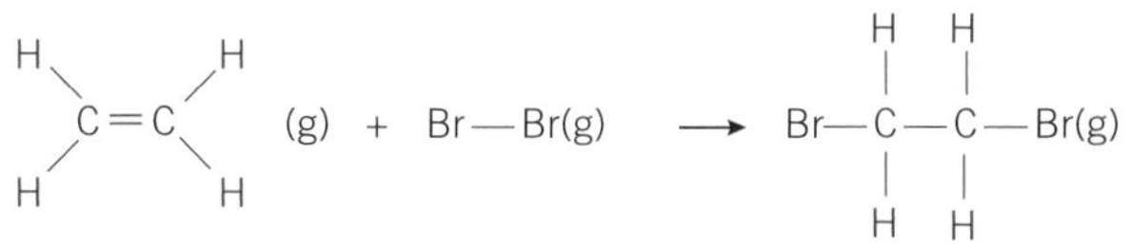

The values for bond enthalpy for a selection of bonds are shown in the margin.

a Why are bond enthalpy values always endothermic? *(1 mark)*

b Calculate the bond enthalpy change for the reaction between ethene and bromine. *(3 marks)*

c What is the significance of the sign for the bond enthalpy change? *(1 mark)*

Bond	Mean bond enthalpy / $kJ\,mol^{-1}$
C=C	+612
C–H	+413
C–Br	+285
Br–Br	+193
C–C	+348

2 A student placed $100\,cm^3$ of $1.00\,mol\,dm^{-3}$ copper(II) sulfate solution into a polystyrene cup and added an excess of magnesium powder.

The equation for the reaction is shown below.

$Mg(s) + CuSO_4(aq) \rightarrow MgSO_4(aq) + Cu(s)$

The temperature of the solution increased from $21.0\,°C$ to $32.0\,°C$

The specific heat capacity of the solution, $c = 4.18\,J\,g^{-1}\,K^{-1}$

The density of all the solutions $= 1.00\,g\,cm^{-3}$

a Why did the student add magnesium powder? *(1 mark)*

b Define the term *specific heat capacity*. *(2 marks)*

c Calculate the energy change in the reaction. *(2 marks)*

d Calculate the amount, in mol, of copper(II) sulfate used in the reaction. *(1 mark)*

e Calculate the enthalpy change of the reaction. Give your answer to three significant figures. *(2 marks)*

f What is the major source of error in this experiment? What could the student do to reduce the effect of this problem? *(2 marks)*

3 During combustion $2.00\,g$ of ethanol, C_2H_5OH heated $250\,cm^3$ of water by $30\,°C$.

The specific heat capacity of the solution, $c = 4.18\,J\,g^{-1}\,K^{-1}$

The density of water $= 1.00\,g\,cm^{-3}$

a Write an equation to represent the complete combustion of ethanol. Include state symbols. *(2 marks)*

b Calculate the heat energy change for the reaction. *(2 marks)*

c Calculate the amount, in mol, of ethanol burnt in the reaction. Give your answer to three significant figures. *(2 marks)*

d Calculate the enthalpy change of the reaction. Give your answer to three significant figures. *(2 marks)*

4 Consider the equation below which shows the enthalpy change of formation of butane, C_4H_{10}.

$4C(s) + 5H_2(g) \rightarrow C_4H_{10}(g)$

a State Hess's law. *(1 mark)*

b Suggest why it would not be possible to measure the enthalpy change of formation of butane directly. *(1 mark)*

5.1 Collision theory
5.2 Maxwell–Boltzmann distribution

Specification reference: 3.1.5

Key term

Activation energy E_a: The minimum collision energy that particles must have to react.

- Different reactions have different activation energies.
- The lower the activation energy the larger the number of particles that can react at any temperature.

Collision theory

The rate of a chemical reaction is a measure of how quickly a reactant is used up or of how quickly a product is made.

You can use the collision theory to understand how the conditions used affect the rate of a chemical reaction.

- For a reaction to occur particles must collide.
- When the particles collide they must have enough energy to break the existing bonds.
- When the particles collide they must collide in the correct orientation so that the reactive parts of the molecules come together. This is particularly important for large molecules. As a result only a small proportion of collisions between particles result in a reaction.

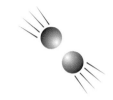

▲ **Figure 1** *Particles must collide before they can react.*

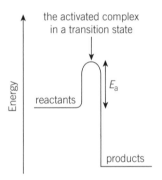

▲ **Figure 2** *The energy change in an exothermic reaction*

The effect of changing the temperature

- As you increase the temperature of a sample, you increase the kinetic energy of the particles.
- As the temperature increases, the particles move faster so they collide more often.
- Also when the particles do collide more of the particles will have enough energy to react. As a result the higher the temperature the larger the number of particles that can react.

However, the particles will only react if they collide in the correct orientation.

Maxwell–Boltzmann distributions

These diagrams are used to represent the energy of the particles in a sample of a gas at a given temperature.

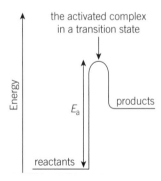

▲ **Figure 3** *The energy change in an endothermic reaction*

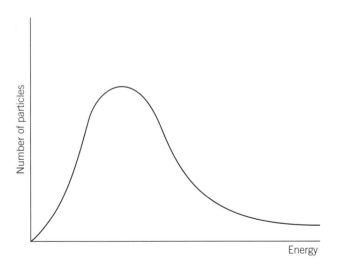

▲ **Figure 4** *The energy distribution curve for a sample of gas at a particular temperature.*

Activation energy and Maxwell–Boltzmann distributions

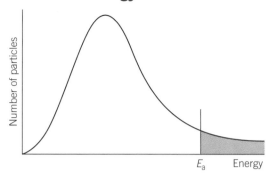

▲ **Figure 5** *Activation energy*

The activation energy is the minimum collision energy that particles must have to react.

- For a reaction to occur the particles must collide and these collisions must have energy equal to or greater than the activation energy. If there is not enough energy when the particles collide then the particles will not react.

- Only the small proportion of particles in the shaded part of the distribution curve have enough energy to react.

- Notice that the activation energy is drawn towards the right of the peak of the distribution curve. If more than half the particles had enough energy to react, the reaction would be too fast to be controlled safely.

Maxwell–Boltzmann distributions and temperature

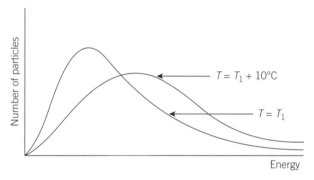

▲ **Figure 6** *The Maxwell–Boltzmann distributions at different temperatures*

Summary questions

1 What is the activation energy of a reaction? *(1 mark)*

2 How does increasing the temperature affect the energy of the molecules? *(1 mark)*

3 Describe how the Maxwell–Boltzmann distribution of energies diagram changes if the temperature is decreased by 10 °C. *(2 marks)*

4 Why does increasing the temperature increase the rate of a chemical reaction? *(2 marks)*

5.3 Catalysts

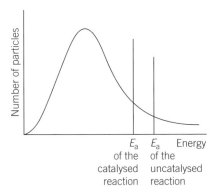

▲ **Figure 1** *A catalysed reaction has a lower activation energy*

Catalysts

- A catalyst increases the rate of a chemical reaction but is not used up itself during the reaction.
- Catalysts work by providing an alternative reaction pathway which has lower activation energy. As a result at any given temperature more particles will have energy greater than or equal to the activation energy. As a result the reactions happen faster (the rate of reaction increases).

Catalysts and Maxwell–Boltzmann distributions

> **Worked example**
>
> Q The graph below represents the Maxwell–Boltzmann distribution of energies at a particular temperature.
>
>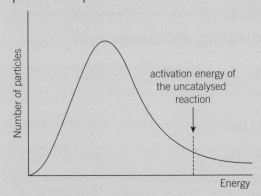
>
> State the effect of adding a catalyst on:
>
> a the energy of the particles
>
> b the activation energy
>
> c the rate of reaction.
>
> A a Adding a catalyst will not affect the energy of the particles.
>
> b Adding a catalyst will provide an alternative reaction pathway with a lower activation energy.
>
>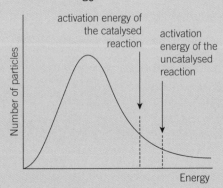
>
> c The rate of reaction will increase because more particles now have enough energy to react.

Chapter 5 Practice questions

1   Maxwell–Boltzmann distribution diagrams can be used to show the distribution of energy of the molecules in a sample of a gas.

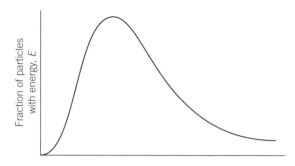

   a Label the *x*-axis of the Maxwell–Boltzmann graph.   (*1 mark*)

   b The temperature of the sample of gas is increased by 10 °C. Draw a new line on the graph to show the distribution of energy of the molecules at this higher temperature.   (*2 marks*)

   c Explain, in terms of the particles, why increasing the temperature of the sample of a gas can increase the rate of a chemical reaction.   (*3 marks*)

2   Consider the Maxwell–Boltzmann distribution diagram below.

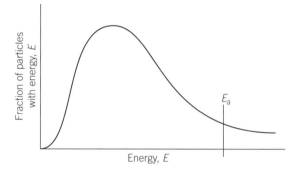

   a Chemical reactions take place when particles collide. Explain, in terms of the particles involved, two reasons why most collisions do not result in a chemical reaction taking place.   (*2 marks*)

   b On the diagram above draw a line to show the effect of adding a catalyst on the activation energy of the reaction.   (*2 marks*)

   c Catalysts can increase the rate of a chemical reaction. Explain how a catalyst can increase the rate of a chemical reaction. Use your diagram to help you explain your answer.   (*2 marks*)

3   **a** Explain the term *collision theory*.   (*2 marks*)

   b Suggest two ways that the collision frequency of the molecules in a gaseous sample could be increased.   (*2 marks*)

6.1 The idea of equilibrium
6.2 changing the conditions of an equilibrium reaction

Specification reference: 3.1.6

Reversible reactions and dynamic equilibria

Many reactions are reversible; they can go forwards or backwards. For example:

$$SO_2(g) + \frac{1}{2}O_2(g) \rightleftharpoons SO_3(g)$$

- The $\rightleftharpoons$ sign means that the reaction is reversible.
- If the forward reaction is exothermic the enthalpy change will have a negative sign. The reverse reaction will be endothermic and the enthalpy change will have a positive sign.
- If a reversible reaction takes place in a closed system where nothing can enter or leave then an equilibrium can be reached.
- At equilibrium there is a balance between the reactants and the products and the concentrations of the products and reactants stay the same. This does not mean that the concentration of reactants is equal to the concentration of the products.
- The forward and backward reactions have not stopped. It is just that at equilibrium the rate of forward reaction is equal to the rate of backward reaction. As a result the equilibrium is described as being dynamic.

Le Chatelier's principle

Le Chatelier's principle helps us to predict the effect of changing factors that affect the position of equilibrium.

The effect of changing the concentration on the position of equilibrium

If a reaction involving solutions is at equilibrium then changing the concentration will affect the position of equilibrium. The position of equilibrium shifts to minimise the change. If the concentration of one of the reactants is increased it will shift the position of equilibrium towards the right so more product will be made.

The effect of changing temperature on the position of equilibrium

If a reaction is at equilibrium then changing the temperature may affect the position of equilibrium. The position of equilibrium will shift to minimise the change in temperature. As a result if the temperature is increased the position of equilibrium will shift in the endothermic direction.

The effect of changing the pressure on the position of equilibrium

If a reaction that involves gases is at equilibrium then changing the pressure may affect the position of equilibrium. The position of equilibrium shifts to minimise the change. If the pressure is increased it will shift the position of equilibrium towards the side that has fewer gas molecules. If the pressure is decreased it will shift the position towards the side with more gas molecules.

Summary questions

1 What does the sign $\rightleftharpoons$ mean? (*1 mark*)

2 What does Le Chatelier's principle state? (*1 mark*)

3 How does adding a catalyst to a chemical reaction affect the rate of reaction and the position of equilibrium? Explain your answers. (*2 marks*)

6.3 Equilibrium reactions in industry
Specification reference: 3.1.6

Equilibrium conditions

The following conditions apply to all chemical equilibria:

- The reactants and products must be in a closed system (where nothing can enter or leave).

- Equilibrium can be reached from either direction (you can start with reactants or products and the same equilibrium position will be reached provided the same conditions are used).

- The equilibrium is dynamic meaning reactants are reacting to form new products and products are reacting together to form the reactants again but the rate of forwards reaction is equal to the rate of backwards reaction.

Equilibria and chemical processes
The Haber process

Ammonia, NH_3, is produced by the Haber process. The reaction is reversible and the forward reaction is exothermic.

$$N_2(g) + 3H_2(g) \rightleftharpoons 2NH_3(g)$$

- An iron catalyst is used to increase the rate of reaction. This allows a reaction to be carried out at a reasonable temperature.

- The catalyst does not affect the position of equilibrium, so it has no effect on the yield of ammonia at equilibrium.

- The forward reaction between nitrogen and hydrogen is exothermic. Increasing the temperature increases the rate of reaction but decreases the yield of ammonia (increasing the temperature shifts the position of equilibrium in the endothermic direction). As a result a compromise temperature is used. This gives a reasonable rate and a reasonable yield of ammonia.

- Increasing the pressure increases the yield of ammonia (increasing the pressure shifts the position of equilibrium towards the products side, which has fewer gas particles).

The Contact process

Sulfuric acid is produced by the Contact process. The reaction is reversible and the forward reaction is exothermic.

$$SO_2(g) + \frac{1}{2}O_2(g) \rightleftharpoons SO_3(g)$$

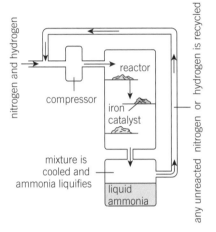

▲ **Figure 1** *The Haber process*

Summary questions

1 What is a closed system? *(1 mark)*

2 What is the catalyst used in the Haber process and how does it affect the rate of reaction and the yield of ammonia made? *(3 marks)*

6.4 The equilibrium constant, K_c

Key term

Chemical equilibrium: When the rate of forward reaction is equal to the rate of backward reaction so that the concentration of the reactants and products remain constant.

Maths skill

For the reaction

$$H_2(g) + I_2(g) \rightleftharpoons 2HI(g)$$

$$K_c = \frac{(mol\,dm^{-3})^2}{(mol\,dm^{-3})\,(mol\,dm^{-3})}$$

= no units

The units cancel out so there are no units

Worked example:

Q For the reaction

$$N_2O_4(g) \rightleftharpoons 2NO_2(g)$$

Write the expression for the equilibrium constant, K_c and state its units.

A $K_c = \dfrac{[NO_2(g)]^2}{[N_2O_4(g)]}$

Units $= \dfrac{(mol\,dm^{-3})\,(mol\,dm^{-3})}{(mol\,dm^{-3})}$

$= mol\,dm^{-3}$

Summary questions

1 Write the expression for the equilibrium constant, K_c for
   a $2NO_2(g) \rightleftharpoons 2NO(g) + O_2(g)$
   b $2HCl(g) \rightleftharpoons H_2(g) + Cl_2(g)$
   c $2SO_2(g) + O_2(g) \rightleftharpoons 2SO_3(g)$ *(3 marks)*

2 State the units for the equilibrium constant, K_c for
   a $2NO_2(g) \rightarrow 2NO(g) + O_2(g)$
   b $2HCl(g) \rightarrow H_2(g) + Cl_2(g)$
   c $2SO_2(g) + O_2(g) \rightarrow 2SO_3(g)$ *(3 marks)*

Reversible reactions

Many reactions are reversible and do not go to completion. Eventually, a mixture that contains both reactants and products is formed. The symbol $\rightleftharpoons$ is used to indicate that the reaction is reversible. If a reaction is carried out in a closed system, equilibrium may be reached.

Homogeneous equilibrium

A homogeneous equilibrium is when all the reactants and products in the equilibrium mixture are in the same physical state or phase, for example, all the reactants and products are gases.

The equilibrium constant, K_c

At equilibrium the concentration of the reactants and of products remains constant.

Hydrogen, H_2, and iodine, I_2, can react together to form hydrogen iodide, HI. However, the reaction is reversible and hydrogen iodide molecules can decompose to form hydrogen and iodine. These reactions can be summed up by the equation below:

$$H_2(g) + I_2(g) \rightleftharpoons 2HI(g)$$

Eventually equilibrium is reached in which the concentration of the two reactants and of the products remains constant.

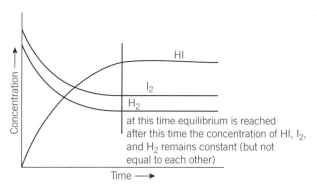

▲ **Figure 1** *Reaching equilibrium*

At equilibrium the concentrations of hydrogen, iodine, and hydrogen iodide are constant and are linked together by the equilibrium constant, K_c.

For the reaction

$$H_2(g) + I_2(g) \rightleftharpoons 2HI(g)$$

$$K_c = \frac{[HI(g)]^2}{[H_2(g)]\,[I_2(g)]}$$

The square brackets in the expression indicate that the concentrations are measured in units of $mol\,dm^{-3}$

Units for K_c

The units for K_c vary and have to be worked out for each equilibrium system.

6.5 Calculations using equilibrium constant expressions

Specification reference: 3.1.6

Calculating the equilibrium constant 1

If the concentration of the reactants and products present at equilibrium are known the value for the equilibrium constant, K_c, can be calculated quite easily by substituting the values of concentration into the expression for the equilibrium constant.

Calculating the equilibrium constant 2

In some questions the number of moles of reactants and products and the total volume of the equilibrium mixture are given. K_c can still be calculated but first, the concentration of the reactants and products must be deduced using the expression:

Concentration (mol dm^{-3}) = amount (mol) / volume (dm^3). Then the values for the concentration can be substituted into the expression for the equilibrium constant, K_c.

Calculating the equilibrium constant 3

The hardest examples occur when the initial amounts of the reactants are given together with the equilibrium amount of one substance. Although K_c can still be found, several steps are required. First the equilibrium amount of each substance must be deduced. Then the equilibrium concentrations must be calculated. Finally the equilibrium concentrations can be substituted into the expression for the equilibrium constant, K_c.

Worked example:

Q $N_2(g)$, $H_2(g)$ and $NH_3(g)$ exist in equilibrium in a closed system.

$$N_2(g) + 3H_2(g) \rightleftharpoons 2NH_3(g)$$

The concentration of the reactants and products were found to be:

$N_2(g) = 0.0200$ mol dm^{-3}

$H_2(g) = 0.0100$ mol dm^{-3}

$NH_3(g) = 0.0800$ mol dm^{-3}

Calculate the value of the equilibrium constant. Include the units in your answer.

A $K_c = \dfrac{[NH_3(g)]^2}{[N_2(g)][H_2(g)]^3}$

$= \dfrac{(0.0800 \times 0.0800)}{0.0200 \times (0.0100 \times 0.0100 \times 0.0100)}$

$= \dfrac{0.0064}{2.00 \times 10^{-8}}$

$= 320\,000$ mol^{-2} dm^6

Summary questions

1 $H_2(g)$, $Cl_2(g)$ and $HCl(g)$ exist in equilibrium in a closed system. The total volume is 1.00 dm^3.

   $$H_{2(g)} + Cl_{2(g)} \rightleftharpoons 2HCl(g)$$

   The concentration of the reactants and products at equilibrium were found to be:

   $H_{2(g)} = 0.0500$ mol dm^{-3}
   $Cl_{2(g)} = 0.100$ mol dm^{-3}
   $HCl_{(g)} = 0.800$ mol dm^{-3}

   Calculate the value of K_c. Give your answer to three significant figures and include the units.   (3 marks)

2 Consider the equation below.

   $$N_2O_4(g) \rightleftharpoons 2NO_2(g)$$

   A chemist allowed 0.600 moles of N_2O_4 to decompose in a 500 cm^3 flask.

   The equilibrium mixture was analysed and found to contain 0.250 moles of N_2O_4

   Calculate the value of K_c. Give your answer to three significant figures and include the units.   (5 marks)

6.6 The effect of changing conditions on equilibria

Specification reference: 3.1.6

Synoptic link

Look back at Topic 5.1, Collision theory, to revise activation energy for reactions.

The equilibrium constant can also be calculated for gases using partial pressures. This equilibrium constant has the symbol K_p.

Le Chatelier's principle ⚠

If a reversible reaction is carried out in a closed system eventually an equilibrium can be reached. However, if the conditions are changed the system is no longer in equilibrium and the position of equilibrium can be shifted.

Changes in pressure

If a reaction that involves gases, and is in dynamic equilibrium, is subject to a change in pressure the position of equilibrium will shift so as to minimise the increase in pressure.

> **Worked example:**
>
> Q Many industrial processes involve reversible reactions.
>
> The Contact process is used in the manufacture of sulfuric acid. In the Contact process SO_2 reacts with O_2 to produce SO_3.
>
> $$2SO_2(g) + O_2(g) \rightleftharpoons 2SO_3(g)$$
>
> State and explain the effect of increasing the total pressure on the position of equilibrium.
>
> A There are 3 gaseous molecules on the left-hand side of the equation and 2 gaseous molecules on the right-hand side of the equation.
>
> Increasing the total pressure will shift the position of equilibrium to the right-hand side as it is the side with the fewer gaseous molecules and will decrease the total pressure.

Changes in concentration

If a reaction in dynamic equilibrium is subject to a change in concentration of reactants or products the position of equilibrium will shift so as to minimise the change in concentration.

Changes in temperature

If a reaction in dynamic equilibrium is subject to a change in temperature the position of equilibrium will shift so as to minimise the change in temperature.

> **Worked example:**
>
> Q Consider the equation below
>
> $$N_2(g) + 3H_2(g) \rightleftharpoons 2NH_3(g) \quad \Delta H = -92\,kJ\,mol^{-1}$$
>
> State and explain the effect of decreasing the temperature on the position of equilibrium.
>
> A The forward reaction is exothermic, whilst the backward reaction is endothermic.
>
> Decreasing the temperature will shift the position of equilibrium to the right-hand side as this is the exothermic direction.

The equilibrium constant, K_c

If a reversible reaction is exothermic in the forward direction, increasing the temperature will decrease the equilibrium constant, K_c. If a reversible reaction is endothermic in the forward direction increasing the temperature will increase the equilibrium constant, K_c.

Summary questions

1. Consider the equation below
   $$N_2(g) + 3H_2(g) \rightleftharpoons 2NH_3(g)$$
   State and explain the effect of a decrease in total pressure on the position of equilibrium.
   (2 marks)

2. Consider the equation below
   $$X(g) + 3Y(g) \rightleftharpoons 2Z(g),$$
   $$\Delta H = -100\,kJ\,mol^{-1}$$
   State and explain the effect of increasing the temperature on the position of equilibrium.
   (2 marks)

3. Consider the equation below
   $$E(g) + 2F(g) \rightleftharpoons G(g)$$
   $$\Delta H = -120\,kJ\,mol^{-1}$$
   State and explain the effect of increasing the temperature on the value of the equilibrium constant, K_c.
   (2 marks)

1 Nitrogen and hydrogen react together to form ammonia.

$N_2(g) + 3H_2(g) \rightleftharpoons 2NH_3(g)$, $\Delta H = -92\,kJ\,mol^{-1}$

A chemist mixes 3.00 moles of nitrogen with 6.00 moles of hydrogen in a 1.00 dm³ flask. At equilibrium there was found to be 2.00 moles of NH_3.

  a Deduce the equilibrium concentrations of nitrogen and hydrogen. *(2 marks)*

  b Write the expression for the equilibrium constant, K_c, for this equilibrium. *(1 mark)*

  c Calculate the value of the equilibrium constant. Give your answer to an appropriate number of significant figures and include the units. *(3 marks)*

  d A chemist increases the temperature of the reaction. State and explain what happens to the value for the equilibrium constant as the temperature is increased. *(1 mark)*

  e A chemist increases the pressure of the reaction mixture. State and explain what happens to the position of equilibrium as the pressure is increased. *(1 mark)*

2 The reaction between sulfur dioxide and oxygen to produce sulfur trioxide is reversible. The equation for the reaction is shown below.

$2SO_2(g) + O_2(g) \rightleftharpoons 2SO_3(g)$

The forward reaction is exothermic.

The system was allowed to reach equilibrium in a closed system.

  a Define the term *closed system*. *(1 mark)*

  b The equilibrium can be described as being dynamic. What can be said about the rate of forward reaction in a dynamic equilibrium? *(1 mark)*

  c State and explain the effect of increasing the temperature on the position of equilibrium. *(2 marks)*

  d Vanadium (V) oxide can be used to catalyse the reaction between sulfur dioxide and oxygen. State and explain the effect of adding a vanadium (V) oxide catalyst on the position of equilibrium. *(2 marks)*

3 Ammonia can be produced from nitrogen and hydrogen using the Haber process. The equation for the reaction is shown below.

$N_2(g) + H_2(g) \rightleftharpoons 2NH_3(g)$

The forward reaction is exothermic.

The reaction is carried out at a temperature of 450 °C and a pressure of 200 atmospheres. A finely divided iron catalyst was used.

  a State and explain the effect of increasing the temperature on: *(2 marks)*

    i the rate of reaction

    ii the position of equilibrium.

  b State and explain the effect of increasing the pressure on: *(2 marks)*

    i the rate of reaction

    ii the position of equilibrium.

  c State and explain the effect of using the iron catalyst on: *(2 marks)*

    i The rate of reaction

    ii The position of equilibrium.

7.1 Oxidation and reduction
7.2 Oxidation states

Specification reference: 3.1.7

Oxygen and hydrogen

The term *redox* is short for reduction–oxidation and these two reactions always occur together. Historically the term *oxidation* was used for reactions in which oxygen is added to a substance and the term reduction was used for reactions in which oxygen was removed from a substance.

Consider the equation below:

$$PbO + H_2 \rightarrow Pb + H_2O$$

Oxygen is added to the hydrogen, H_2, to form water, H_2O, so the hydrogen is oxidised. At the same time oxygen is removed from the lead(II) oxide, PbO to form lead, Pb so the lead(II) oxide is reduced. In a similar way historically the term *reduction* was used for reactions in which hydrogen was added to a substance.

$$CH_3CHO + 2[H] \rightarrow CH_3CH_2OH$$

Hydrogen, H, is added to the ethanal, CH_3CHO, to form ethanol, CH_3CH_2OH, so the ethanal is reduced. If the reaction were reversed and hydrogen was removed from ethanol to form ethanal the ethanol would be oxidised.

Oxidation and reduction

As one species loses electrons another species must gain these electrons. As a result oxidation and reduction must always occur together. These reactions are often called 'redox reactions'.

Oxidising and reducing agents

- An oxidising agent is a species that oxidises another substance by removing electrons from it.
- Oxidising agents are themselves reduced during the reaction.
- Common oxidising agents include oxygen, chlorine, potassium dichromate(VI), and potassium manganate(VII).
- A reducing agent is a species that reduces another substance by adding electrons to it.
- Reducing agents are themselves oxidised during the reaction.
- Common reducing agents include Group 1 and 2 metals, hydrogen, carbon, and carbon monoxide.

Assigning oxidation states

- Any pure element has an oxidation state of zero.
- Any monatomic ion has an oxidation state equal to the charge on the ion.
- In compounds, Group 1 metal atoms have an oxidation state of +1.
- In compounds, Group 2 metal atoms have an oxidation state of +2.
- In compounds, the most electronegative element fluorine always has an oxidation state of −1.

- In compounds, hydrogen has an oxidation state of +1 unless it is in a metal hydride when it has an oxidation state of −1.
  - The total oxidation state of an uncharged molecule is 0.
  - The total oxidation state of any polyatomc ion is equal to the charge of the ion.

Examples

The oxidation state of hydrogen in H_2O is +1.

The oxidation state of hydrogen in KH is −1.

- In compounds, the oxidation state of oxygen is −2 unless it is with fluorine when it has an oxidation state of +2 or is part of peroxide when it has an oxidation state of −1.

Examples

The oxidation state of oxygen in H_2O is −2.

The oxidation state of oxygen in F_2O is +2.

The oxidation state of oxygen in H_2O_2 is −1.

- If a molecule is neutral overall, for example CO_2, then the sum of the oxidation states of the atoms in the molecule must be zero.
- Where a molecular ion has an overall charge, for example CO_3^{2-}, then the sum of the oxidation states of the atoms in the molecule must equal the overall charge of the ion.

Summary questions

1 What is the oxidation state of oxygen in O_2?   (1 mark)

2 What is the oxidation state of sulfur in SO_4^{2-}?   (1 mark)

3 State the oxidation number of iodine in:
   a KIO_3 b KI   c KIO_4 (3 marks)

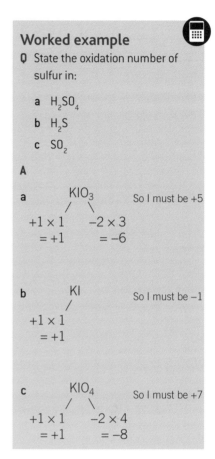

Worked example

Q State the oxidation number of sulfur in:

   a H_2SO_4

   b H_2S

   c SO_2

A

a KIO_3 So I must be +5

   $+1 \times 1$ -2×3
   $= +1$ $= -6$

b   KI   So I must be −1

   $+1 \times 1$
   $= +1$

c KIO_4 So I must be +7

   $+1 \times 1$ -2×4
   $= +1$ $= -8$

Maths skill

Make sure you state the oxidation state of each atom. For example in $K_2Cr_2O_6$ the oxidation state of Cr is +6.

7.3 Redox equations

Specification reference: 3.1.7

Half-equations

The reaction for the oxidation of magnesium can be summed up by the equation

$$2Mg + O_2 \rightarrow 2MgO$$

Half-equations can be used to show what happens to each of the reactants in a redox reaction.

$$2Mg \rightarrow 2Mg^{2+}$$

$$O_2 + 4e^- \rightarrow 2O^{2-}$$

Using oxidation states to identify oxidation and reduction reactions

When magnesium is added to copper sulfate solution a displacement reaction takes place.

The overall equation for the reaction is

$$Mg + CuSO_4 \rightarrow MgSO_4 + Cu$$

- The oxidation state of uncombined magnesium is zero; in $MgSO_4$ it is +2. The oxidation state has gone up so the magnesium is oxidised.

- The oxidation state of copper is +2 in $CuSO_4$ and zero in uncombined copper. The oxidation state has gone down so the copper has been reduced.

- We can show the electron transfer by splitting the overall equation into two half-equations.

$$Mg \rightarrow Mg^{2+} + 2e^- \text{ (oxidation)}$$

$$Cu^{2+} + 2e^- \rightarrow Cu \text{ (reduction)}$$

The magnesium is oxidised and the copper is reduced. The magnesium has reduced the copper so it is a reducing agent. The copper sulfate has oxidised the magnesium so it is an oxidising agent. Take care to identify the name of the reagent rather than just the copper as the oxidising agent. The sulfate ions are not involved in the reaction and are referred to as being spectator ions.

Summary questions

1 Define the terms *oxidation* and *reduction*.   (*2 marks*)

2 Consider the two half-equations below.
   $Fe^{2+} + 2e^- \rightarrow Fe$
   $Zn \rightarrow Zn^{2+} + 2e^-$
   Combine these two half-equations to give the overall equation
   for the reaction.   (*1 mark*)

3 Consider the equation below:
   $Zn + 2HCl \rightarrow ZnCl_2 + H_2$
   Split the overall equation into two half-equations.   (*2 marks*)
4 Magnesium reacts with chlorine to form magnesium chloride.
   a Write the overall equation for the reaction.
   b Give the two half-equations for the reaction.   (*3 marks*)

Chapter 7 Practice questions

1 Iron is a transition element and forms ions with several different oxidation states.

    a Deduce the oxidation state of the iron in each of these substances. *(2 marks)*

        i $FeCl_3$

        ii $FeCO_3$

    b Give the full name of each of these compounds. *(2 marks)*

        i $FeCl_3$

        ii $FeCO_3$

2 Complete the table below to show the oxidation state of sulfur in each of the substances.

Substance	Oxidation state of S
H_2S	
SO_2	
SO_3	
H_2SO_4	
S_8	

(5 marks)

3 Consider the reaction below

$$Fe(s) + CuSO_4(aq) \rightarrow Cu(s) + FeSO_4(aq)$$

    a What is the oxidation state of the reagent iron? *(1 mark)*

    b Use oxidation states to explain which species are oxidised and reduced in this reaction. *(2 marks)*

    c Name the oxidising agent in this reaction. *(1 mark)*

4 Magnesium reacts with hydrochloric acid in a redox reaction.

The equation for the reaction is shown below.

$$Mg(s) + 2HCl(aq) \rightarrow MgCl_2(aq) + H_2(g)$$

    a Describe what you would see during this reaction. *(2 marks)*

    b Explain, in terms of electron transfer, which species is oxidised and which is reduced in this reaction. *(2 marks)*

5 Complete the table below to show the oxidation state of the element required in each substance.

	Oxidation state
Mg in $MgCO_3$	
Cu in Cu_2O	
Cl in $NaClO_4$	

(3 marks)

6 Which of these statements describes an oxidising agent?

    A A substance that is oxidised.

    B An element that is oxidised.

    C A substance that gains electrons during a chemical reaction

    D A substance that does not change oxidation state during a chemical reaction. *(1 mark)*

7 Identify the oxidation state of Cl in the compound $NaClO_3$

    A −1   **C** −5

    B +5   **D** +7 *(1 mark)*

8.1 The Periodic Table
8.2 Trends in the properties of elements of Period 3

Specification reference: 3.2.1

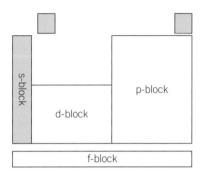

▲ **Figure 1** *Blocks of the Periodic Table*

Synoptic link

To revisit covalent bonding, look back at Topic 3.2, Covalent bonding.

Synoptic link

To revisit metallic bonding, look back at Topic 3.3, Metallic bonding.

Synoptic link

To revisit intermolecular bonding, look back at Topic 3.5, Forces acting between molecules.

Key term

Periodic: Recurs regularly.

s-block, p-block, and d-block elements

Elements that belong to the s-block have their highest energy electrons in s-orbitals.

For example, lithium, Li $1s^2 2s^1$, is an s-block element.

p-block elements have their highest energy electrons in p-orbitals

For example, nitrogen, N $1s^2 2s^2 2p^3$ is a p-block element.

d-block elements

d-block elements have their highest energy electrons in d-orbitals.

For example, titanium, Ti $1s^2 2s^2 2p^6 3s^2 3p^6 4s^2 3d^2$ is a d-block element.

Groups

A group is a vertical column in the Periodic Table. Elements in the same group have the same number of electrons in their outer shell and so have similar chemical properties.

Periods

A period is a horizontal row in the Periodic Table.

Periodicity

Examining the Periodic Table reveals many patterns in the properties of the elements.

Across each period the metals are found on the left and the non-metals on the right. This is a regularly recurring pattern and is an example of periodicity.

Melting points and boiling points

Trends in melting point and boiling point across Period 3 are related to the change in bonding and structure of the elements.

▼ **Table 1** *Table of the melting points and boiling points of Period 3 elements*

Element	Melting point/K	Boiling point/K
Na	371	1156
Mg	922	1380
Al	933	2740
Si	1683	2628
P	317	553
S	392	718
Cl	172	238
Ar	84	87

Sodium, magnesium, and aluminum are metals and have a giant metallic structure. Metallic bonds are strong so their melting points and boiling points are quite high.

Silicon is a semi-metal, which forms a giant covalent (macromolecular) structure in which each silicon atom is bonded to four other silicon atoms by very strong covalent bonds so the melting point and boiling points of silicon are very high.

Phosphorus, P_4, sulfur, S_8, and chlorine, Cl_2, are non-metals that form simple molecules. The van der Waals' bonds between these molecules are weak so these elements melt and boil at low temperatures.

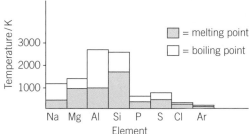

▲ **Figure 2** *Melting points and boiling points of the Period 3 elements*

▼ **Table 2** *Period 3 elements*

Element	Type of bonding structure	Force broken on melting
sodium	giant metallic	metallic bond
magnesium	giant metallic	metallic bond
aluminium	giant metallic	metallic bond
silicon	giant covalent	covalent bond
phosphorus	simple covalent, P_4 units	van der Waals' force
sulphur	simple covalent, S_8 units	van der Waals' force
chlorine	simple covalent, Cl_2 units	van der Waals' force
argon	monatomic	van der Waals' force

More about structures of some Period 3 elements

Sodium, magnesium, and aluminum are all metals. In magnesium each atom has two electrons in its outer shell. Both these electrons are delocalised leaving a magnesium ion with a 2+ charge. The metallic bonds in magnesium are stronger than the metallic bonds in sodium (with only one delocalised electron and ions with a 1+ charge) so the melting point of magnesium is higher than the melting point of sodium.

The bonding in silicon is giant covalent with a network of covalent bonds throughout the structure. Melting or boiling silicon requires the breaking of covalent bonds so requires a lot of energy.

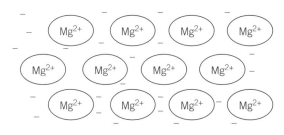

▲ **Figure 3** *The metallic bonding in magnesium*

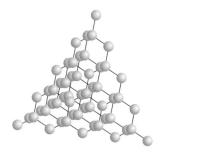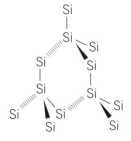

▲ **Figure 4** *Silicon has a giant covalent structure*

Phosphorus, sulfur, and chlorine are all simple molecules with van der Waals' forces between them. This force strength increases with the increasing size of the molecule (number of electrons) so the melting or boiling point increases in the order argon, chlorine, phosphorus, and sulfur.

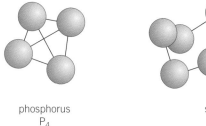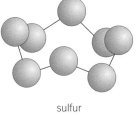

phosphorus P_4 sulfur S_8 chlorine Cl_2

▲ **Figure 5** *Phosphorus, sulfur, and chlorine have a simple molecular structure*

8.3 More trends in the properties of elements of Period 3
8.4 A closer look at ionisation energies

Specification reference: 3.2.1

Key term

First ionisation energy: First ionisation energy is the energy required to remove one electron from each atom in one mole of gaseous atoms forming one mole of ions with a single positive charge.

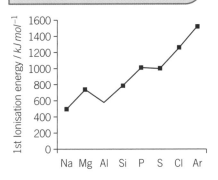

▲ **Figure 1** First ionisation energy for Period 3 elements

Summary questions

1 Define the term *first ionisation energy*. (*2 marks*)

2 State and explain the trend in the atomic radii of Period 3 elements. (*2 marks*)

3 Give the equation that represents the first ionisation energy of magnesium. (*2 marks*)

4 Why does sulfur have a lower first ionisation energy than phosphorus? (*2 marks*)

Atomic radius

The **atomic radius** of the elements decreases across Period 3 as the number of protons in the nucleus increases across the period but the corresponding extra electrons are placed into the same shell. This means that the electrostatic attraction of the positive nucleus on the negative electrons increases slightly across the period, making the atomic radius slightly smaller.

Electron configurations and first ionisation energy

It is important that you can describe and explain the graph showing the first ionisation energy for the elements in Period 3. There is a general increase across Period 3 as the number of protons in the nucleus steadily increases, leading to a greater charge on the nucleus. As a result there is a greater attraction between the nucleus and the outer electron so more energy is required to remove the electron from the gaseous atom.

Why are the increases in ionisation energy not regular?

There is a slight decrease between magnesium and aluminium because the outer electron in aluminium is in a p sub-level which is higher in energy than the outer electron in magnesium which is in an s sub-level. So aluminium loses its outer electron slightly more easily than aluminium loses its outer electron.

▼ **Table 1** *The electron configurations of magnesium and aluminium*

Element	Electron configuration	Spin diagrams					
		1s	2s	2p	3s	3p
Mg	$1s^2\,2s^2\,2p^6\,3s^2$	⇅	⇅	⇅ ⇅ ⇅	⇅	
Al	$1s^2\,2s^2\,2p^6\,3s^2\,3p^1$	⇅	⇅	⇅ ⇅ ⇅	⇅	↑

There is another slight decrease between phosphorus and sulfur. The outer electron configuration of sulfur has a pair of electrons in one of the p-orbitals. There is repulsion between the electrons so less energy is needed to remove an electron from this pair so sulfur has a lower first ionisation energy than phosphorus.

▼ **Table 2** *The electron configurations of phosphorus and sulfur*

Element	Electron configuration	Spin diagrams					
		1s	2s	2p	3s	3p
P	$1s^2\,2s^2\,2p^6\,3s^2\,3p^3$	⇅	⇅	⇅ ⇅ ⇅	⇅	↑ ↑ ↑
S	$1s^2\,2s^2\,2p^6\,3s^2\,3p^4$	⇅	⇅	⇅ ⇅ ⇅	⇅	⇅ ↑ ↑

Chapter 8 Practice questions

1   Complete the table below to show the bonding and structure of the Period 3 elements.

Element	Bonding	Structure
Na		
Mg	metallic	
Al		giant metallic lattice
Si	covalent	
P$_4$		simple molecules
S$_8$	covalent	
Cl$_2$		
Ar	van der Waals'	Monatomic

(2 marks)

2   Sodium, aluminium, and chlorine belong to Period 3 of the Periodic Table.

   a   In terms of their structure and bonding, explain why chlorine has a lower melting point than sodium.   (3 marks)

   b   In terms of their structure and bonding explain why aluminium has a higher melting point than sodium.   (2 marks)

3   Phosphorus, P$_4$, sulfur, S$_8$, and chlorine, Cl$_2$, exist as simple molecules. Predict which of these elements has the highest melting point. Explain your answer.   (3 marks)

4   Identify the Period 3 element that has the following successive ionisation energies. Explain your answer.   (1 mark)

	1st	2nd	3rd	4th	5th	6th
Ionisation energy / kJ mol^{-1}	577	1820	2740	11600	14800	18400

5   In terms of structure and bonding, explain why silicon has a very high melting point.   (3 marks)

6   Which Period 3 element has the second largest atomic radius?

   A   Na

   B   Ar

   C   Cl

   D   Mg   (1 mark)

Go Further

1   Give the equation that represents the first ionisation energy of nitrogen.
   (2 marks)

2   Explain why oxygen has a lower first ionisation energy than nitrogen.
   (3 marks)

9.1 The physical and chemical properties of Group 2

Specification reference: 3.2.2

Element	Atomic radius / nm
Mg	0.145
Ca	0.194
Sr	0.219
Ba	0.253

Element	First ionisation energy / $kJ\,mol^{-1}$
Mg	738
Ca	590
Sr	550
Ba	503

Revision tip

Notice how in figure 1 the calcium ions have a 2+ charge.

Introducing the alkaline earth metals

Group 2 metals have two electrons in their outer shell. They react to form ions that have a 2+ charge.

$$Mg \rightarrow Mg^{2+} + 2e^-$$

Trend in atomic radius

Down the group atoms have an extra shell of electrons and become larger. As a result the atomic radii of the elements increase down the group.

Trend in ionisation energy

- Down the group the outer electrons are lost more easily.
- Although the number of protons increases down the group, the atomic radii also increase so the distance between the nucleus and the outer electrons increases.
- There is also more shielding by the inner electrons. As a result the value for the first ionisation energy decreases down the group.

Trend in melting points

Group 2 elements have reasonably high melting points due to metallic bonding.

Element	melting point / $^\circ C$
Mg	649
Ca	839
Sr	769
Ba	729

Metallic bonding

In metals the electrons in the outer shell of atoms are delocalised. This leads to positive metal ions and negatively charged delocalised electrons.

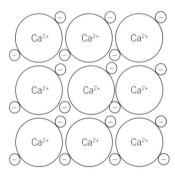

▲ **Figure 1** *The metallic bonding in calcium metal*

- Metallic bonding is the electrostatic attraction between the positive ions and the negative delocalised electrons.
- Down the group the size of the metal ions increases so the strength of the metallic bonding decreases.
- Less energy is required to overcome the forces of attraction. As a result melting points generally decrease down the group.

Trend in reactivity

- The reactivity increases down the group.
- Group 2 elements are strong reducing agents because they can lose electrons quite easily.
- Down the group the elements become stronger reducing agents.

How do Group 2 elements react with water?

Group 2 metal atoms are good reducing agents. Their reaction with water is an example of a redox reaction.

- The Group 2 metal atom is oxidised.

$$M \rightarrow M^{2+} + 2e^-$$

- The hydrogen in water is reduced.

Magnesium reacts slowly with water to form magnesium hydroxide and hydrogen.

$$Mg(s) + 2H_2O(l) \rightarrow Mg(OH)_2(aq) + H_2(g)$$

Magnesium hydroxide is sparingly soluble.

Magnesium reacts quickly with steam.

$$Mg(s) + H_2O(g) \rightarrow MgO(s) + H_2(g)$$

Calcium, strontium, and barium all react vigorously with water to form a metal hydroxide and hydrogen.

Example

$$Ca(s) + 2H_2O(l) \rightarrow Ca(OH)_2(aq) + H_2(g)$$

A solution of calcium hydroxide looks cloudy as it is only slightly soluble in water and the undissolved calcium hydroxide forms a white suspension.

Trend in the solubilities of the Group 2 hydroxides

The solubility of Group 2 hydroxides increases down the group.

Group 2 hydroxide	Solubility (g per 100 cm³ of water)	Solubility
Mg(OH)₂	0.0012	↓
Ca(OH)₂	0.113	
Sr(OH)₂	0.410	
Ba(OH)₂	2.57	

Trend in the solubilities of the Group 2 sulfates

The solubility of Group 2 sulfates decreases down the group.

Group 2 sulfate	Solubility (g per 100 cm³ of water)	Solubility
MgSO₄	25.5	↑
CaSO₄	0.24	
SrSO₄	0.013	
BaSO₄	0.00022	

Summary questions

1 State the trend in the solubilities of the Group 2 sulfates. *(1 mark)*

2 State and explain the trend in the atomic radii of the Group 2 elements. *(1 mark)*

3 State and explain the general trend in the melting points of the Group 2 elements. *(1 mark)*

51

Chapter 9 Practice questions

1 State and explain the change in the first ionisation energy of Group 2 elements. (*4 marks*)

2 A student added an excess of magnesium powder to a solution of hydrochloric acid.

The equation for the reaction is shown below:

$$Mg(g) + 2HCl(aq) \rightarrow MgCl_2(aq) + H_2(g)$$

State and explain which species are oxidised and reduced in this reaction. (*2 marks*)

3 Choose one answer in each section. (*3 marks*)

   a Down Group 2 the size of the atoms:

   increases/decreases/stays the same.

   b Down Group 2 the melting point of the elements:

   increases/decreases/stays the same.

   c Down Group 2 the strength of the metallic bonding in the elements:

   increases/decreases/stays the same.

4 Magnesium hydroxide is sometimes called milk of magnesia. Give a medical use of magnesium hydroxide and explain why it works. (*2 marks*)

5 Give the equation for the second ionisation energy of strontium. (*2 marks*)

6 Give the full electron configuration of: (*3 marks*)

   a Mg

   b Ca

   c Ca^{2+}

7 Which of these Group 2 metals has the second smallest atoms? (*1 mark*)

   A Be

   B Mg

   C Sr

   D Ba

8 Which equation represents the first ionisation energy of magnesium? (*1 mark*)

   A $Mg \rightarrow Mg^+ + e^-$

   B $Mg(g) \rightarrow Mg(g)^+ + e^-$

   C $Mg^+ + e^- \rightarrow Mg$

   D $Mg(g) \rightarrow Mg(g)^{2+} + 2e^-$

10.1 The halogens

Specification reference: 3.2.3

Introducing the halogens

The elements in Group 7 of the Periodic Table are often called the halogens.

Halogen atoms have seven electrons in their outer shell.

When halogen atoms react they can gain an electron to form a halide ion which has a –1 charge.

Trend in atomic radius

Down the group atoms have an extra shell of electrons and become larger. As a result the atomic radii (size) of the elements increase down the group.

Trends in electronegativity

Element	Electronegativity
F	4.0
Cl	3.0
Br	2.8
I	2.5

Down the group the atoms have a higher atomic number so the greater number of protons should be able to attract the electrons more. However, down the group:

- The atoms have more shells of electrons. This means there is more shielding and less attraction between the nucleus and the electrons.
- The atomic radius increases. This results in less attraction between the nucleus and the electrons in the covalent bond. As a result down the group the halogens become less electronegative.

Trends in boiling point

At room temperature:

- Fluorine is a yellow gas.
- Chlorine is a pale green gas.
- Bromine is a brown liquid.
- Iodine is a dark grey solid.

Notice how the boiling points of the halogens increase down the group.

- The halogens exist as diatomic molecules, for example, chlorine exists as chlorine molecules, Cl_2.
- This means that there are strong covalent bonds within the halogen molecules but only very much weaker van der Waals' forces of attraction between halogen molecules.
- The strength of the van der Waals' forces depends on the number of electrons in the molecules.
- Down the group the number of electrons in the halogen molecules increases, so the strength of the Van der Waals' forces of attraction between molecules increases. As a result down the group the boiling point increases.

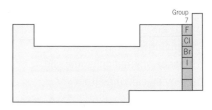

▲ **Figure 1** *Group 7 elements are typical non-metals*

Synoptic link

Look back at Topic 3.2, Covalent bonding.

Synoptic link

Look back at Topic 3.5, Forces acting between molecules.

Key term

Electronegativity: A way of measuring the ability of an atom to draw the electrons in a covalent bond to itself.

Summary questions

1 Define the term *electronegativity*. (*1 mark*)

2 State and explain the trend in atomic radius down Group 7. (*2 marks*)

3 State and explain the trend in electronegativity down Group 7. (*2 marks*)

The chemical reactions of halogens

Key terms

Oxidising agent: A species that oxidises another substance by removing electrons from it.

Oxidation: The loss of electrons.

Reduction: Reduction is the gain of electrons.

Half-equation: A half-equation is either of the two equations that describe each half of a redox reaction.

Trend in the oxidising ability of the halogens

The oxidising ability of a substance is a measure of the strength of an atom to attract and gain an electron.

- The halogens are good oxidising agents.
- Oxidising agents are themselves reduced during the reaction.

When halogen atoms react they gain electrons.

Example

$$Cl_2 + 2e^- \rightarrow 2Cl^-$$

The oxidising ability of the halogens decreases down the group. Down the group the electron that is gained is placed into a shell that is further from the nucleus.

- Down the group the atomic radii increase.
- The amount of shielding increases. As a result the attraction between the nucleus and the electron decreases. As a result the oxidising ability of the halogens decreases down the group. Fluorine is a very powerful oxidising agent.

Displacement reactions

The displacement reactions between halogens and aqueous halides demonstrate the decrease in the oxidising ability of the halogens down the group.

Chlorine is a more powerful oxidising agent than bromine so chlorine oxidises bromide ions.

Example

$$Cl_2(aq) + 2Br^-(aq) \rightarrow 2Cl^-(aq) + Br_2(aq)$$

The half-equations for this reaction are

$$Cl_2(aq) + 2e^- \rightarrow 2Cl^-(aq) \text{ (reduction)}$$

Chlorine is reduced to chloride.

$$2Br^-(aq) \rightarrow + Br_2(aq) + 2e^- \text{(oxidation)}$$

Bromide is oxidised to bromine.

Once the halogen has been displaced it forms a solution. Different halogens form different coloured solutions. These colours can be used to show a reaction has taken place. However, it is difficult to tell bromine and iodine apart when they are dissolved in water. If an organic solvent, such as cyclohexane, is added and the mixture shaken the halogen dissolves in the organic solvent to give a much clearer colour.

▼ **Table 1** *Identifying halogens*

Halogen	Water	Cyclohexane
chlorine, Cl_2	pale green	pale green
bromine, Br_2	orange	orange
iodine, I_2	brown	violet

Summary questions

1 Define the term *oxidising agent*. (*1 mark*)

2 State and explain the trend in the oxidising ability of the halogens. (*3 marks*)

3 A chemist bubbles chlorine, Cl_2 gas through a solution of potassium iodide, KI.
  i Give the equation for the reaction.
  ii State the colour that would be observed when cyclohexane is added shaken with the reaction mixture.
  iii State which species has been oxidised and explain your answer. (*4 marks*)

10.3 Reactions of halide ions
10.4 Uses of chlorine
Specification reference: 3.2.3

Trends in the reducing ability of halide ions

Reducing agents are oxidised during the reaction. The larger the ion the more easily it loses an electron. Down the group the halide ions become increasingly good reducing agents. As a result iodide ions are the strongest reducing agent. Iodide ions are easiest to oxidise.

Identifying halide ions 🧪

You can identify halide ions using acidified silver nitrate solution.

- Fluoride ions form silver fluoride, which is a soluble salt so no precipitate is seen.
- The other halide ions form insoluble salts. The colour of the silver salt can be used to identify the halide.

▼ **Table 1** *Identifying halide ions*

Halide ion	Salt formed	Colour of the precipitate
chloride	AgCl	white
bromide	AgBr	cream
iodide	AgI	yellow

The solubility of the silver halides in ammonia can be used to confirm the identity of the halide ions.

▼ **Table 2** *Confirming the identity of halide ions*

Silver halide	Solubility in ammonia
AgCl	Dissolves in dilute ammonia solution.
AgBr	Does not dissolve in dilute ammonia solution but does dissolve in concentrated ammonia solution.
AgI	Does not dissolve even in concentrated ammonia solution.

Chlorine and water treatments

Chlorine reacts with water to form two acids: chloric(I) acid and hydrogen chloride.

$$Cl_2(aq) + H_2O(l) \rightleftharpoons HClO(aq) + HCl(aq)$$

The reaction of chlorine with sodium hydroxide

Chlorine reacts with cold, dilute sodium hydroxide to form sodium chloride, sodium chlorate(I) NaOCl, and water. This is another example of disproportionation. The chlorine is simultaneously oxidised and reduced. Sodium chlorate(I) is used to make household bleaches.

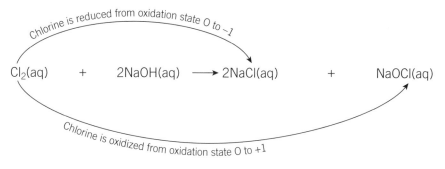

▲ Figure 2

Key term

Reducing agent: A reducing agent is a species which reduces another substance by adding electrons to it.

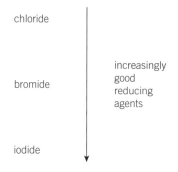

chloride

bromide

iodide

increasingly good reducing agents

▲ **Figure 1**

Revision tip

The oxidation state of chlorine in Cl_2 is 0; in HClO is +1 and in HCl is −1.

This is a disproportionation reaction.

The chlorine is simultaneously oxidised (from 0 to +1) and reduced (from 0 to −1).

Summary questions

1 Give the oxidation number of chlorine in;
   a KCl
   b $NaClO_3$ *(2 marks)*

2 How could you identify bromide ions?   *(2 marks)*

3 State and explain the trend in the reducing ability of the halide ions.   *(2 marks)*

55

1 A student has a sample of a Group 1 halide. The student wants to identify the halide in the sample. Outline what the student could do to identify the halide ion present in the sample. Include any observations that the student should make. *(5 marks)*

2 Choose one answer in each section: *(3 marks)*

   a Down Group 7 the size of the atoms: increases/decreases/stays the same.

   b Down Group 7 the melting point of the elements: increases/decreases/stays the same.

   c Down Group 2 the strength of the van der Waals' bonding between molecules: increases/decreases/stays the same.

3 A halide ion can be displaced by a more reactive halogen. The equation for the reaction between a solution of chlorine and a solution of sodium bromide is shown below.

$$Cl_2(aq) + 2NaBr(aq) \rightarrow Br_2(aq) + 2NaCl(aq)$$

   a What you would observe during this reaction? *(1 mark)*

   b State which species is oxidised and which species is reduced during the reaction. Explain your answer. *(2 marks)*

   c Name the oxidising agent in this reaction. *(1 mark)*

4 Which of these halogens is the least electronegative? *(1 mark)*

   A F

   B Cl

   C Br

   D I

5 Which of the options below correctly identifies the oxidation number of the chlorine in each substance.

	KCl	NaOCl	$NaClO_4$
A	+1	+1	+7
B	−1	−1	+3
C	−1	+1	+7
D	−1	+1	−7

(1 mark)

6 Which of these species is the most powerful reducing agent?

   A F^-

   B Cl^-

   C Br^-

   D I^- *(1 mark)*

11.1 Carbon compounds
Specification reference: 3.3.1

The chemistry of carbon

Organic chemistry is the study of compounds containing carbon combined with other elements. Carbon can form rings and chains that can be branched or unbranched. Carbon atoms have four electrons in their outer shell and can from four covalent bonds to other atoms. In addition carbon bonds are quite strong. The ability to form these bonds means that a very large number of organic compounds exist.

Empirical formula

The empirical formula of a compound can be easily deduced from its molecular formula.

The empirical formula of a compound can also be calculated from percentage composition by mass data.

Worked example

Q An organic compound was analysed and found to contain:

52.2% carbon

34.8% oxygen

13.0% hydrogen

A Calculate the empirical formula of the organic compound.

▼ **Table 1** *Empirical formula = C_2OH_6*

Element	C	O	H
% by mass	52.2	34.8	13.0
Divide by relative atomic mass	$\frac{52.2}{12.0} = 4.35$	$\frac{34.8}{16.0} = 2.175$	$\frac{13.0}{1.0} = 13$
Divide by smallest value	$\frac{4.35}{2.175} = 2$	$\frac{2.175}{2.175} = 1$	$\frac{13}{2.175} = 6$

Molecular formula

The molecular formula of a compound can be easily worked out from its displayed formula.

Worked example

Q What is the molecular formula of this organic compound?

A The compound contains three carbon atoms, six hydrogen atoms and one oxygen atom. The molecular formula of the compound is C_3H_6O.

The molecular formula of a compound is a whole number multiple of its empirical formula. The molecular formula can be calculated using the empirical formula of the compound and the relative molecular mass of the compound.

Key terms

Empirical formula: The simplest whole number ratio of the atoms in a compound.

Molecular formula: The actual number of each type of atom in a compound.

Skeletal formula: These are simplified formula where straight lines are drawn to represent carbon–carbon bonds. Hydrogen atoms joined to the carbon atoms are removed but other functional groups are shown.

Displayed formula: A formula showing all the bonds in a compound.

Structural formula: A formula showing in minimum detail the arrangement of the atoms in the molecule.

Maths skill

Ethene has the molecular formula C_2H_4.

For every carbon atom there are two hydrogen atoms.

Its empirical formula is CH_2.

Displayed formula

The displayed formula is a very useful way of showing the structure of organic compounds as every atom and bond is shown.

A single line – represents a single covalent bond.

A double line = represents a double covalent bond. For example:

The organic compound butane has the molecular formula C_4H_{10}.

Its displayed formula is

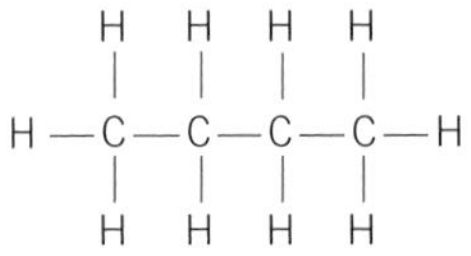

```
      H   H   H   H
      |   |   |   |
  H — C — C — C — C — H
      |   |   |   |
      H   H   H   H
```

Skeletal formula

Organic compounds can have very complicated structures and it can become impractical to draw out displayed formula for such complex structures.

Skeletal formulae are a very useful method of representing the molecules in a simplified but precise way. The carbon skeleton of the compound is shown together with any functional groups. For example:

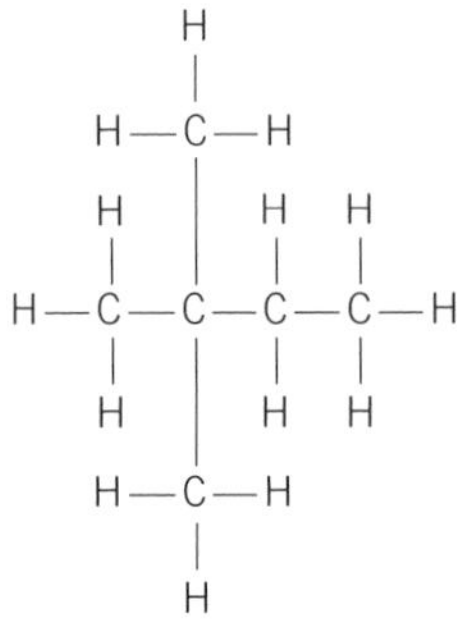

displayed formula 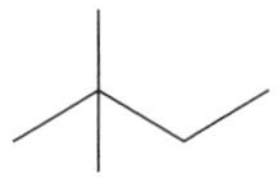   skeletal formula

Structural formula

Molecular formulae can be useful but sometimes two compounds have the same molecular formula. A structural formula is a more useful to chemists. It reveals the structure of the molecule in a very concise way. For example:

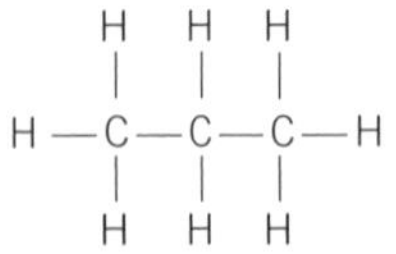

```
      H   H   H
      |   |   |
  H — C — C — C — H
      |   |   |
      H   H   H
```
displayed formula

$CH_3\,CH_2\,CH_3$

structural formula

Summary question

1 Consider the molecule drawn below.

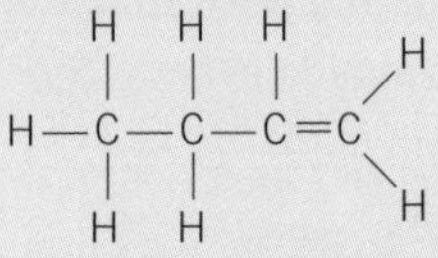

```
      H   H   H      H
      |   |   |     /
  H — C — C — C = C
      |   |         \
      H   H          H
```

a What type of formula is shown in the diagram?
b Give the molecular formula of the compound.
c Give the structural formula of the compound.
d Give the empirical formula of the compound.
e Give the skeletal formula of the compound. (5 marks)

11.2 Nomenclature – naming organic compounds

Specification reference: 3.1.1

Naming organic molecules

▼ **Table 1** *Functional groups*

Functional group	Name
alkane	-ane
alkene	-ene
haloalkane	bromo-, chloro-, iodo-
alcohol	-ol

The alkanes

The alkanes are a homologous series which means they have the same functional group. Alkanes have the general formula C_nH_{2n+2}. For straight-chain alkanes the beginning of the name tells you the number of carbon atoms in the chain.

*   For example methane has one carbon atom in the chain; ethane has two carbon atoms, and so on.

If the chain is branched then it is said to have a side chain. The side chain is named methyl, ethyl, propyl, etc. depending on the number of carbon atoms it contains. This is then put at the front of the name of the main chain.

*   For example methylbutane has a main chain of four carbon atoms and a side chain of one carbon atom.

If there is more than one position that the side chain could attach to the main chain then it is given a number. This is counted from the end of the chain in order to make the number of the side chain as small as possible.

methylbutane
$CH_3CH(CH_3)CH_2CH_3$
C_5H_{12}

2,3-dimethylpentane
$CH_3CH(CH_3)CH(CH_3)CH_2CH_3$
C_7H_{16}

The alkenes

The alkene family is another homologous series. They have a double covalent bond between two of the carbon atoms.

Alkenes have the general formula C_nH_{2n}.

The alkenes are named in the same way as the alkanes except that their name ends in -ene. The position of the C=C double bond is numbered if there is more than one place where it can go in a molecule. The numbering is such that it has the smallest number possible.

▼ **Table 2** *Number of carbon atoms*

Number of C atoms	Prefix
1	meth
2	eth
3	prop
4	but
5	pent
6	hex

▼ **Table 3** *Side chains*

number of C atoms	structure	Name
1	$-CH_3$	methyl
2	$-CH_2CH_3$	ethyl
3	$-CH_2CH_2CH_3$	propyl

▼ **Table 4** *Number of identical side chains*

Number of identical side chains	Prefix
2	di
3	tri
4	tetra

Summary question

1   Name the following compounds:

   a C_6H_{14} (unbranched)
   b C_5H_{12} (unbranched)
   c $CH_3CHCHCH_3$
   d $CH_3CH_2CH(CH_2CH_3)$
       $CH_2CH_2CH_3$
   e $CH_3CHClCH_2CH_3$
   f $CH_3CH_2CH_2CHClCH_3$
       (6 marks)

11.3 Isomerism

Specification reference: 3.3.1

Synoptic link

E–Z isomerism is discussed further in Topic 14.1, Alkenes.

Key terms

Isomerism: Isomerism occurs when two or more organic molecules have the same molecular formula but different arrangements of atoms.

Structural isomerism: Structural isomerism occurs when two or more molecules have the same molecular formula but different structural formulae.

Isomerism

There are two main types of isomerism: structural isomerism and stereoisomerism.

Structural isomerism

There are three types of structural isomerism: positional, functional, and chain.

Positional isomerism

Positional isomerism occurs when a functional group can be in more than one position on the carbon chain.

You must consider positional isomerism when the molecule contains a chain of four or more carbon atoms or when the functional group is a halogen that can be in different positions on the chain.

- In butene (C_4H_8) there are two possible positions for the C=C, between atoms 1 and 2 or between atoms 2 and 3.

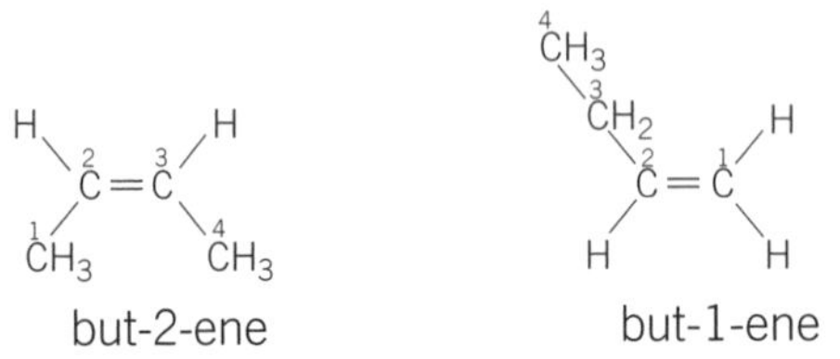

but-2-ene   but-1-ene

▲ **Figure 1** *Positional isomers of C_4H_8*

Functional isomerism

Functional isomerism occurs when there are two different functional groups.

propan-1-ol is an alcohol   ethoxymethane is an ether

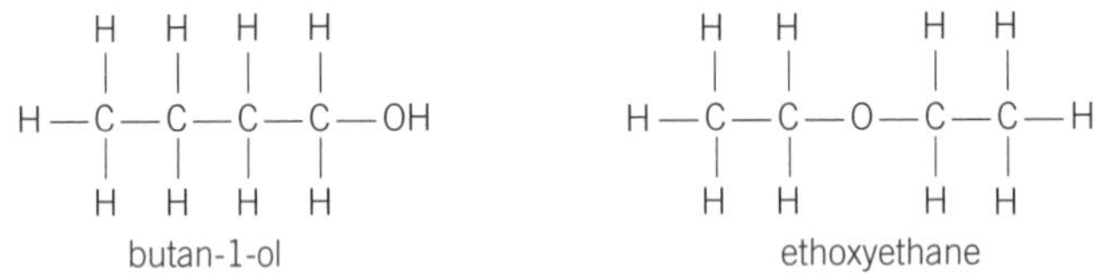

butan-1-ol   ethoxyethane

▲ **Figure 2** *Butan-1-ol and ethoxyethane are functional isomers*

Chain isomerism

Chain isomerism occurs when two or more molecules have the same molecular formula but different arrangements of the carbon chain. Typically this will involve straight chain and branched chain structures.

For example, there are two chain isomers of C_4H_{10} as the carbon atoms can be arranged in a straight chain forming butane or as a chain of three carbon atoms with one side chain forming methylpropane.

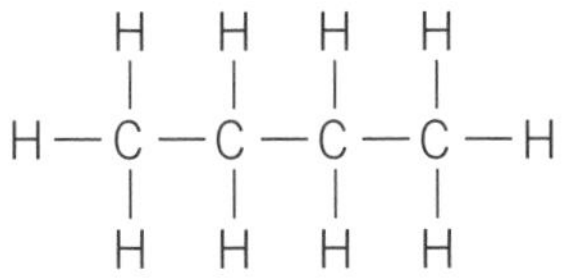

butane

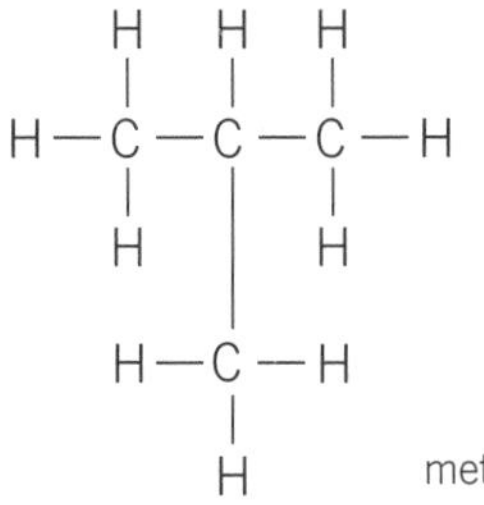

methylpropane

▲ **Figure 3** *Butane and methylpropane are chain isomers*

For C_5H_{12} there are three possible chain isomers; the carbons atoms can be arranged in a straight chain forming pentane, as a chain of four atoms with a side chain forming methylbutane or as a chain of three atoms with two side chains forming dimethylpropane (the prefix di is used to show that there are two methyl groups attached to the central carbon atom).

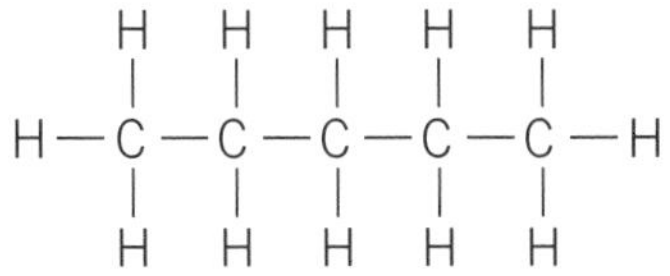

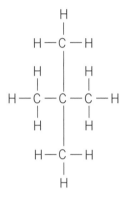

pentane

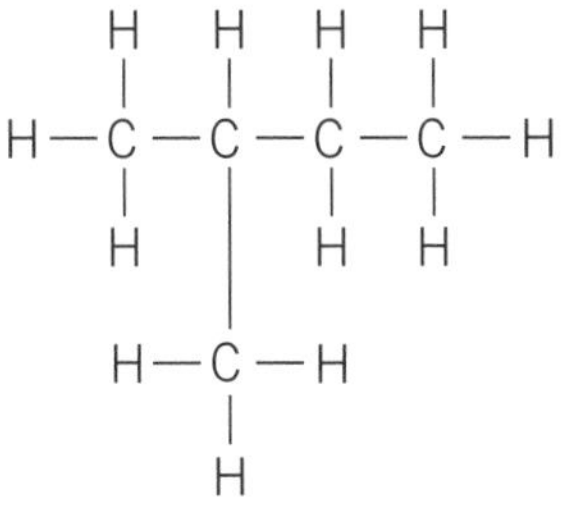

methylbutane

dimethylpropane

▲ **Figure 4** *Chain isomers*

More complex examples

For C_5H_{10} there are five possible isomers. Note that these are a combination of positional and chain isomers.

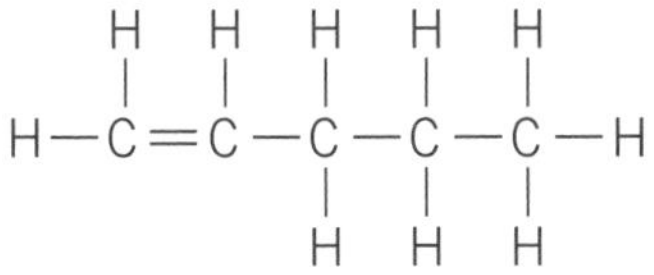

pent-1-ene

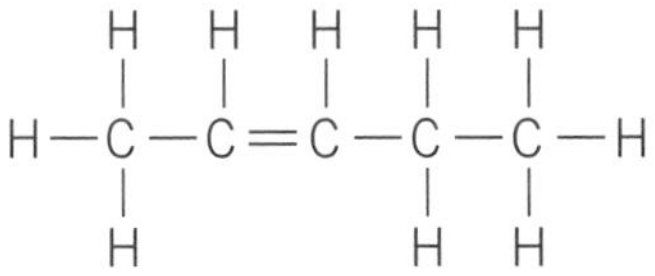

pent-2-ene

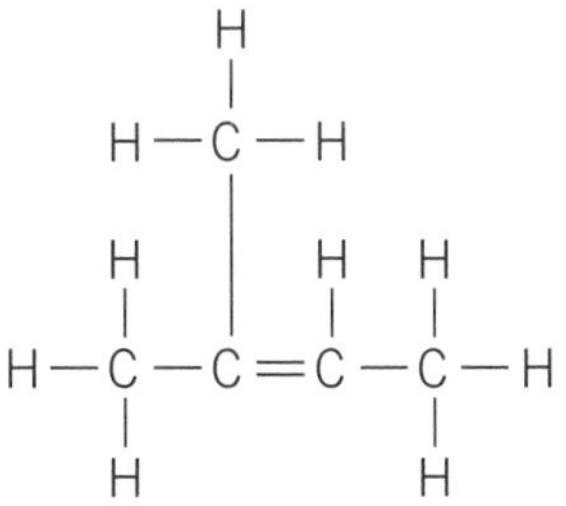

2-methylbut-2-ene

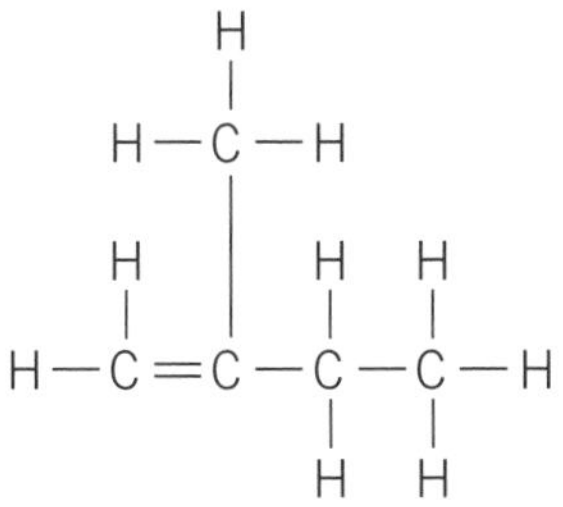

2-methylbut-1-ene

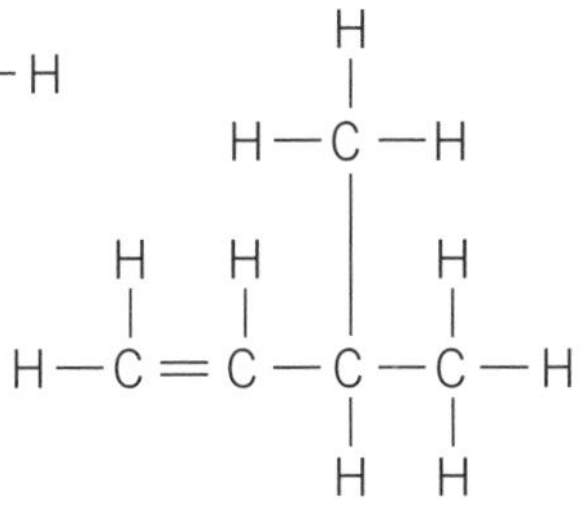

3-methylbut-1-ene

▲ **Figure 5** *Isomers of* C_5H_{10}

Revision tip

It is important that you approach drawing isomers in a systematic way to ensure that you do not miss any out. Name each one when you have drawn it to make sure that you do not draw the same molecule twice.

Summary questions

1. What are the three types of structural isomerism? *(1 mark)*
2. Draw displayed formulae and name all the chain isomers of C_4H_{10}. *(2 marks)*
3. A straight chain organic compound has the molecular formula C_4H_9I. Draw displayed formulae and name the positional isomers of C_4H_9I. *(2 marks)*

Chapter 11 Practice questions

1 An unbranched hydrocarbon **A** has the molecular formula C_4H_8.

  a Define the term *hydrocarbon*. *(1 mark)*

  b Draw the displayed formula and name the three possible structure of hydrocarbon **A**. *(3 marks)*

2 Give the skeletal formula and names of the two organic molecules with the molecular formula C_3H_7Br. *(2 marks)*

3 Which of these options shows the functional group in each of the following organic families? *(1 mark)*

	Alkene	Aldehyde	Carboxylic acid
a	C=C	—C(=O)H	C(=O)(O—H)H
b	C—C	—C(=O)H	—C(=O)OH
c	C=C	—C(=O)H	—C(=O)OH
d	C—C	—C(=O)OH	—C(=O)—

4 The alkanes are a homologous series of hydrocarbons. The table below shows the name and molecular formula of the first five members of the alkane homologous series.

Name	Molecular formula
methane	CH_4
ethane	C_2H_6
propane	C_3H_8
butane	C_4H_{10}
pentane	C_5H_{12}

  a Define the term *homologous series*. *(2 marks)*

  b What is the general formula of the alkane homologous series? *(1 mark)*

  c Give the molecular formula of the tenth member of the alkane homologous series. *(1 mark)*

5 A sample of a hydrocarbon was analysed and found to contain 83.7% carbon. The hydrocarbon has a relative molecular mass of 86.0

  a Define the term *empirical formula*. *(1 mark)*

  b Calculate the empirical formula of the hydrocarbon. *(3 marks)*

  c Deduce the molecular formula of the hydrocarbon. *(1 mark)*

6 Which of these compounds is a member of the alkane homologous series?

  A C_3H_6 **B** C_2H_2

  C C_3H_8 **D** C_2H_4

7 **a** Define the term *structural isomer*. *(1 mark)*

  b Name the positional isomer of propan-1-ol. *(1 mark)*

8 What is the relative molecular mass of butane?

  A 26   **B** 44

  C 56   **D** 58 *(1 mark)*

12.1 Alkanes

Specification reference: 3.3.2

The alkanes

The alkanes are saturated hydrocarbons. This means that alkanes contain single covalent bonds (not double covalent bonds) and they only contain carbon and hydrogen atoms.

For example:

- The alkane with only 1 carbon atom has $(2 \times 1 + 2) = 4$ hydrogen atoms and the molecular formula CH_4.
- The alkane with 26 carbon atoms has $(2 \times 26 + 2) = 54$ hydrogen atoms and the molecular formula $C_{26}H_{54}$.

Naming alkanes

The alkanes all have names ending in -ane. The first part of their name is deduced from the number of carbon atoms in the longest carbon chain.

Examples of straight chain alkanes

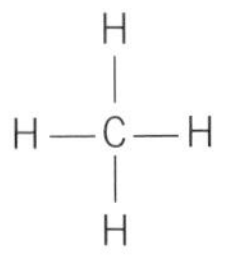

methane
CH_4

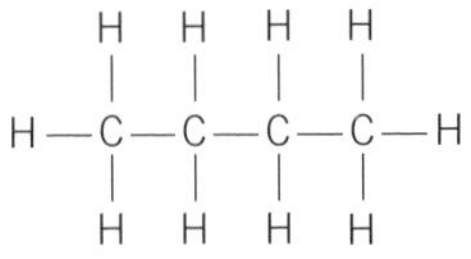

butane
C_4H_{10}

Examples of branched alkanes

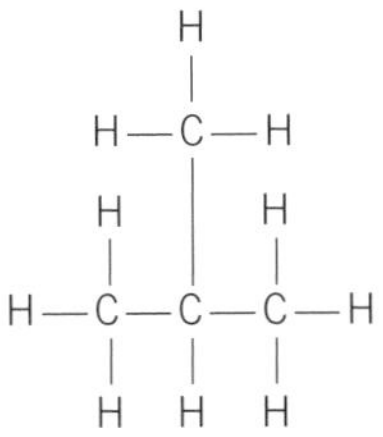

methyl propane
C_4H_{10}

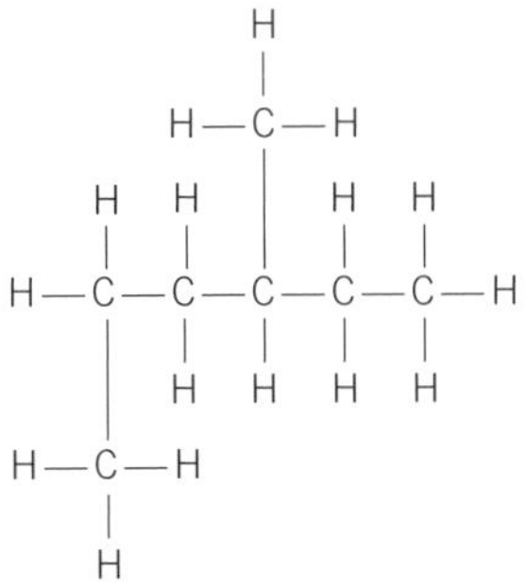

3-methyl-hexane

Examples of ring alkanes

In cyclic or ring alkanes the carbon atoms join together to form a ring of atoms. Ring alkanes have a molecular formula of C_nH_{2n}.

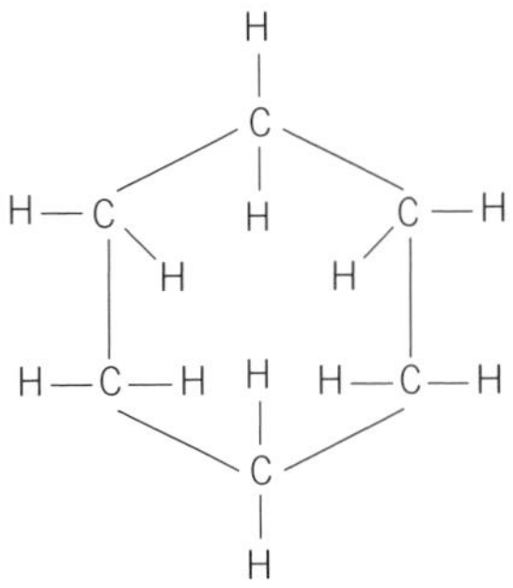

The properties of alkanes
Polarity

There is very little difference in electronegativity between carbon and hydrogen atoms so alkanes are almost non-polar molecules.

cyclohexane
C_6H_{12}

cyclopentane
C_5H_{10}

> **Revision tip**
> Alkanes have the general formula C_nH_{2n+2}

> **Synoptic link**
> See Topic 11.2, Nomenclature – naming organic compounds, for more details on naming alkanes.

> **Revision tip**
> Notice how cyclic alkanes do not have the same general formula as straight chain alkanes.

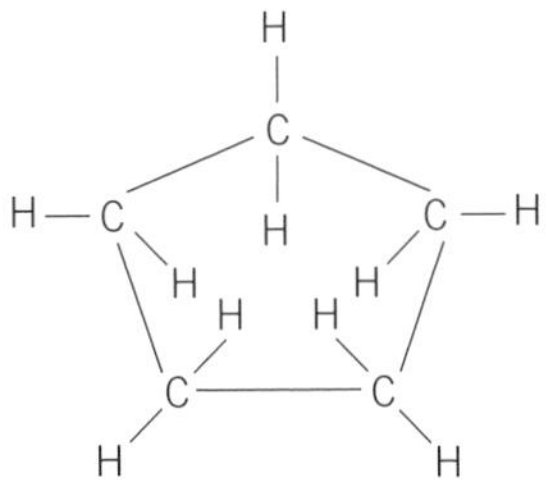

Boiling points 🔬

The boiling points of alkanes increase as the number of carbon atoms in the molecules increases. Short chain alkanes (1 to 4 carbon atoms) are gases at room temperature. Medium chain alkanes (5 to 17 carbon atoms) are liquid at room temperature. Long chain alkanes (18 or more carbon atoms) are solid at room temperature. As the molecules get larger there are more electrons and therefore stronger van der Waals' forces between the molecules so their melting points and boiling points increase.

Solubilty

Alkanes are almost non-polar molecules so they do not dissolve in polar solvents such as water. However, they will mix with non-polar solvents.

Summary questions

1 What is a saturated hydrocarbon? *(2 marks)*

2 What is the general formula of an alkane? *(1 mark)*

3 Name the straight chain hydrocarbons that have the molecular formula:
 a C_2H_6
 b C_3H_8
 c C_5H_{12} *(3 marks)*

4 Name these organic molecules: *(2 marks)*

5 State and explain the trend in the boiling points of the alkane family. *(2 marks)*

6 Why doesn't propane dissolve in water? *(1 mark)*

12.2 Fractional distillation of crude oil

Specification reference: 3.3.2

The origin of crude oil

Crude oil is a very important raw material. It is a fossil fuel that formed millions of years ago from the fossilised remains of dead plants and animals that were buried at high temperatures and pressures below the Earth's surface. Crude oil is a mixture of many substances but mainly consists of branched and unbranched alkane molecules. Small amounts of other substances, such as sulfur, can also be found in crude oil. When crude oils that contain sulfur are burnt sulfur dioxide can be formed. This gas can form acid rain.

Fractional distillation

Crude oil is a mixture and it must be separated into different parts or fractions. Crude oil is a very important source of organic chemicals and this is done on an industrial scale by fractional distillation.

In this process, alkanes are separated into groups with similar boiling points. These groups are called fractions.

The fractionating tower

- The petroleum is heated to around 350 °C so that it forms a vapour.
- The vapour is then passed into the fractionating tower.
- Note that the temperature of the tower falls as it is ascended (this is called a negative temperature gradient).
- The largest hydrocarbons have the highest boiling points so remain as liquids at the bottom of the tower.
- These are tapped off as residue.
- The smallest hydrocarbons have the lowest boiling points so rise up through the tower and leave the top as gases.
- The remaining hydrocarbons rise up the tower until they reach the region that corresponds to their boiling point.
- They then condense at the bubble caps and are removed from the tower as liquids.

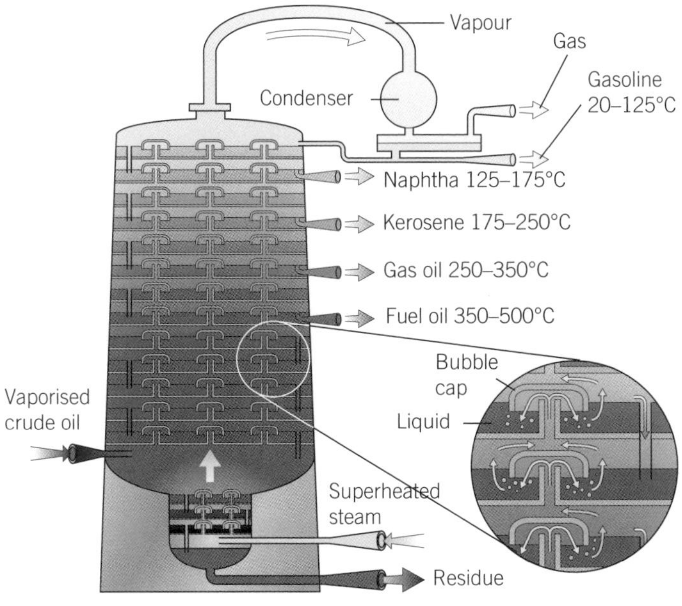

> **Key term**
>
> **Fractional distillation:** The process of separating the crude oil is called fractional distillation. This relies on the differences in boiling points of the different alkane molecules.

▲ **Figure 1** *The fractionating tower*

> **Summary questions**
>
> 1 Describe how crude oil is formed. (*1 mark*)
>
> 2 What is the environmental problem associated with fuels that contain sulfur? (*1 mark*)
>
> 3 In fractional distillation what is a *fraction*? (*1 mark*)
>
> 4 Explain how crude oil is separated into different fractions. (*3 marks*)

12.3 Industrial cracking

Specification reference: 3.3.2

Key term

Cracking: A decomposition reaction which involves the breaking of C—C single bonds.

Maths skill

Make sure that cracking equations balance correctly.

What is cracking?

The fractional distillation of crude oil produces a range of fractions with different boiling points. The mixture of fractions obtained often contains a higher proportion of heavier fractions than is needed. In industry chemists crack these heavier fractions in order to make lighter, more useful ones. Cracking results in the formation of shorter chain alkanes and alkenes.

The heptane molecule can be cracked in a number of ways. In each equation note that both an alkane and an alkene are formed.

$$C_7H_{16} \rightarrow C_5H_{12} + C_2H_4$$

$$C_7H_{16} \rightarrow C_4H_{10} + C_3H_6$$

$$C_7H_{16} \rightarrow C_3H_8 + 2C_2H_4$$

$$C_7H_{16} \rightarrow C_5H_{10} + C_2H_4 + H_2$$

The shorter chain alkanes are more useful as fuels than the original alkane as they are more volatile (evaporate more readily) and burn more easily. The alkenes are used as raw materials for making polymers such as poly(ethene).

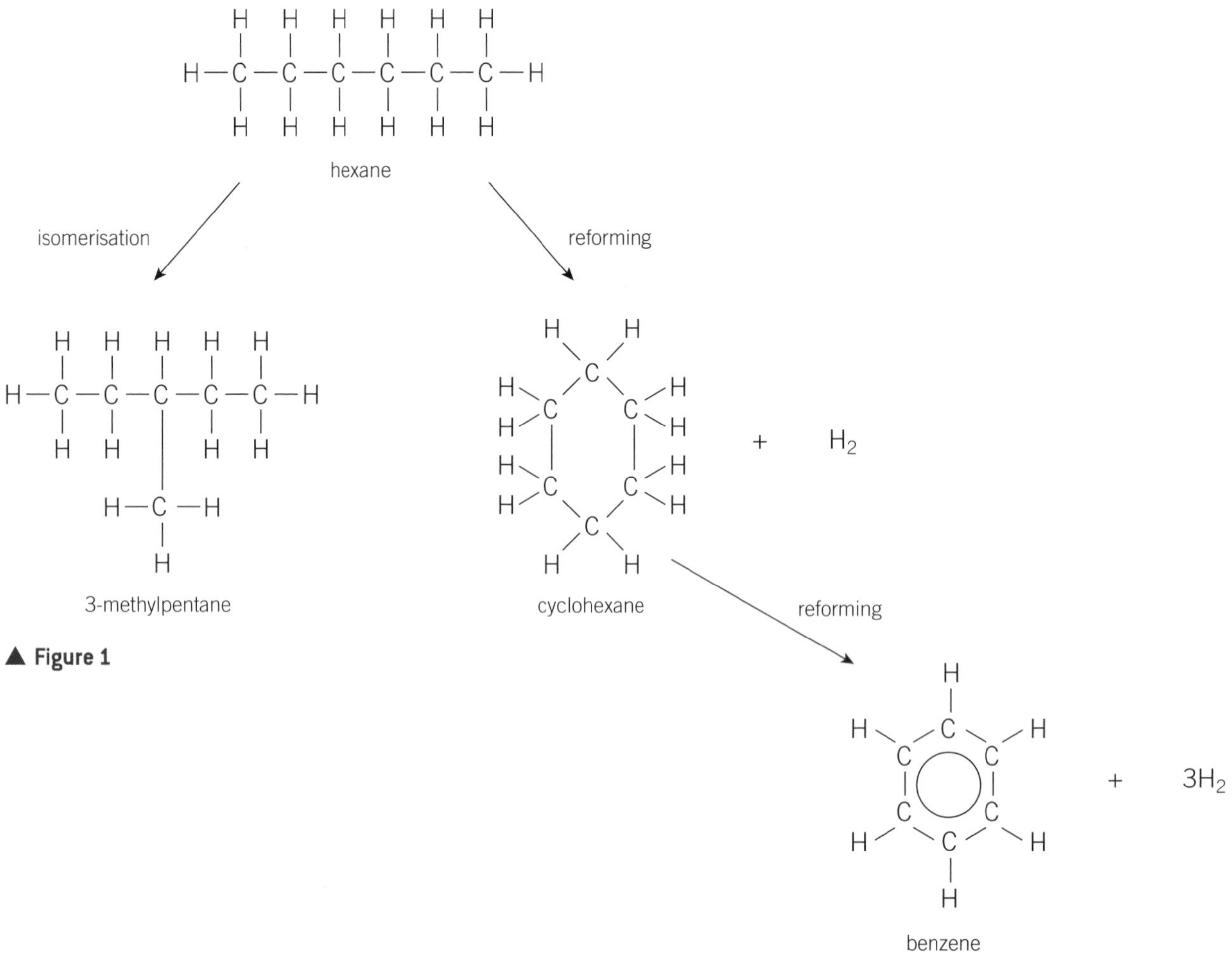

▲ Figure 1

66

Thermal and catalytic cracking

There are two ways cracking can be carried out: thermal cracking and catalytic cracking.

You need to be aware of the differences in reaction conditions used to carry out the two types of cracking.

	Thermal cracking	Catalytic cracking
raw material	long chain alkane	long chain alkane
temperature	800–900 °C	500 °C
pressure	up to 7000 kPa	slightly above atmospheric pressure
catalyst	none	silica and aluminium oxide or zeolite
products	alkene, short chain alkane	aromatic hydrocarbons
uses of products	making polymers	motor fuels (short chains burn more readily, more volatile)
notes	• air is excluded from this process • heating carried out for <1 second to prevent total decomposition.	• more efficient than thermal cracking • produces more branched, cyclic, and aromatic hydrocarbons

Economic reasons for cracking

The fractional distillation process produces disproportionate amounts of some of the fractions of crude oil.

- The amount of petrol produced is not enough for our needs.
- Too much of the naphtha fraction is produced.

Revision tip

The cracking process helps us make maximum use of crude oil by converting the less useful fractions into more useful ones. This is an essential process, as petroleum is a non-renewable resource.

Summary questions

1 Complete the equations below to show the cracking of some alkane molecules.
   a $C_8H_{18} \rightarrow C_5H_{12} + \underline{\quad}$
   b $C_9H_{20} \rightarrow \underline{\quad} + C_2H_4$
   c $C_{22}H_{46} \rightarrow \underline{\quad} + C_4H_8$ (3 marks)

2 Write equations using displayed formulae for the cracking of:
   a Hexane, C_6H_{14} where one of the products is ethene.
   b Octane, C_8H_{18} where one of the products is propene.   (2 marks)

12.4 Combustion of alkanes

Specification reference: 3.3.2

Good fuels 🌡️

Fuels are substances that can be burnt to release heat energy. Short chain alkanes are good fuels which burn completely in a good supply of air to produce carbon dioxide and water and they release a lot of heat energy. Combustion reactions are always exothermic (have a negative sign).

Types of combustion
Complete combustion

For complete combustion to occur:

- A large excess of oxygen is needed.
- alkane + oxygen $\rightarrow$ carbon dioxide + water (+ release of energy)

For methane (the main fossil fuel in natural gas):

$$CH_4(g) + 2O_2(g) \rightarrow CO_2(g) + 2H_2O(l)$$

For octane, a longer chain hydrocarbon, the products are identical.

$$C_8H_{18}(g) + 12\tfrac{1}{2}O_2(g) \rightarrow 8CO_2(g) + 9H_2O(l)$$

Incomplete combustion

In a *limited supply* of oxygen, incomplete combustion occurs.

- The hydrogen in the hydrocarbon still forms water.
- The carbon only undergoes partial oxidation to form carbon monoxide.
- In some cases unburnt carbon particles are released as soot during incomplete combustion.

The incomplete combustion of the same fuels as shown above gives:

$$CH_4(g) + 1\tfrac{1}{2}O_2(g) \rightarrow CO(g) + 2H_2O(l)$$

$$C_8H_{18}(g) + 8\tfrac{1}{2}O_2(g) \rightarrow 8CO(g) + 9H_2O(l)$$

- Carbon monoxide is a toxic gas that binds to the haemoglobin in red blood cells and prevents them carrying oxygen.
- Carbon monoxide is difficult to detect because it is colourless, odourless, and tasteless.

Environmental problems

When alkanes are burnt in engines a lot of heat energy is released and high temperatures can be reached. At these high temperatures enough energy is available for the nitrogen and the oxygen in air to react together forming a number of oxides of nitrogen. These are referred to in general as NO_x. One of the gases formed is nitrogen monoxide which can undergo further oxidation in the air to nitrogen dioxide.

$$N_2(g) + O_2(g) \rightarrow 2NO(g)$$

$$2NO(g) + O_2(g) \rightarrow 2NO_2(g)$$

Some of the fuel passes through the car engine without undergoing oxidation. These gases are called unburned hydrocarbons or volatile organic compounds.

In sunlight these unburned hydrocarbons compounds can react with NO_x to form photochemical smog.

Acid rain

Our rain is naturally acidic because it contains dissolved carbon dioxide from the atmosphere. We use the term acid rain for rain that is more acidic than normal. Acid rain may be formed naturally (from volcanic emissions) or by human activity.

Nitrogen dioxide formed in engines forms nitric acid:

$$2NO_2(g) + H_2O(l) + \tfrac{1}{2}O_2(g) \rightarrow 2HNO_3(aq)$$

Fossil fuels contain sulfur, which forms sulfur dioxide during combustion. This then dissolves in water forming sulfurous acid:

$$SO_2(g) + H_2O(l) \rightarrow H_2SO_3(aq)$$

A series of reactions then lead to the formation of sulfuric acid, $H_2SO_4(aq)$. Acid rain can cause chemical weathering and acidification.

Tackling environmental problems
The catalytic converter

A catalytic converter is fitted to the exhaust system of a car and converts harmful gases into less polluting gases. It removes pollutants from the exhaust gases before they are released through the end of the exhaust. The converter contains a honeycomb mesh of two catalysts (platinum–rhodium and platinum–palladium).

At the platinum–rhodium catalyst the NO_x is reduced to nitrogen.

$$2NO(g) \rightarrow N_2(g) + O_2(g)$$

At the platinum–palladium catalyst two oxidation reactions occur. Any carbon monoxide formed is oxidised to carbon dioxide.

$$CO(g) + \tfrac{1}{2}O_2(g) \rightarrow CO_2(g)$$

Unburned hydrocarbons are oxidised to carbon dioxide and water.

Flue-gas desulfurisation

Flue gas is the waste gas from boilers and furnaces that may contain sulfur dioxide. Flue-gas desulfurisation is the process that removes sulfur dioxide from waste gases. It is used in coal-fired power stations, which release large volumes of sulfur dioxide. A number of reactions occur during the flue-gas desulfurisation process. The key equation is that for the reaction of calcium oxide, CaO, with sulfur dioxide gas forming calcium sulfite:

$$CaO(s) + SO_2(g) \rightarrow CaSO_3(s)$$

The calcium sulfite formed is almost insoluble in water and presents a disposal problem. It can, however, be oxidised to make hydrated calcium sulfate which is gypsum, used to make plasterboard:

$$CaSO_3(s) + \tfrac{1}{2}O_2(g) + 2H_2O(l) \rightarrow CaSO_4 \bullet 2H_2O(s)$$

Synoptic link

You learnt about catalysts in Topic 5.3, Catalysts.

Summary questions

1 Define the term *fuel*. (1 mark)

2 Define the term *complete combustion*. (1 mark)

3 Write balanced equations for the complete combustion of butane. (2 marks)

4 Explain how sulfur in fuels can cause environmental problems. (1 mark)

12.5 The formation of halogenoalkanes

Specification reference: 3.3.2

Key term

Free radical: A species with an unpaired electron.

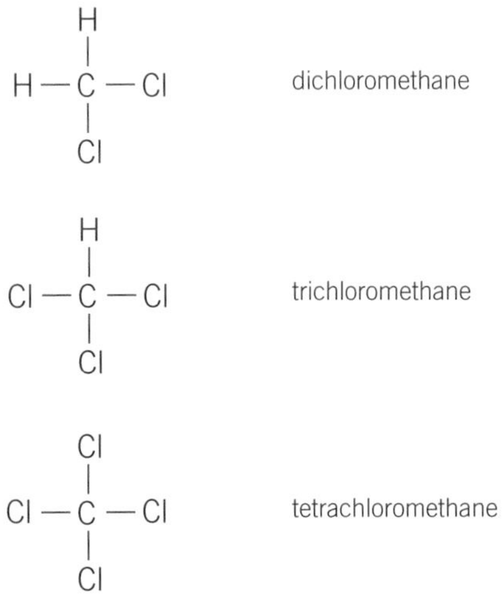

$$H - \underset{\underset{Cl}{|}}{\overset{\overset{H}{|}}{C}} - Cl \qquad \text{dichloromethane}$$

$$Cl - \underset{\underset{Cl}{|}}{\overset{\overset{H}{|}}{C}} - Cl \qquad \text{trichloromethane}$$

$$Cl - \underset{\underset{Cl}{|}}{\overset{\overset{Cl}{|}}{C}} - Cl \qquad \text{tetrachloromethane}$$

▲ **Figure 1** *Products of further substitution reactions*

Revision tip

Many old refrigerators and freezers contain CFCs. When they are disposed of the CFCs must first be removed to prevent the gases escaping into the atmosphere.

Summary questions

1   Define the term 'free radical'.   (*1 mark*)

2   Name the type of reaction by which methane reacts with chlorine.   (*2 marks*)

3   Ethane can react with chlorine to produce chloroethane.
    a   State the conditions required for the reaction.
    b   Write an equation for:
        i   the initiation step
        ii   the propagation steps
        iii a possible termination step.
                (*5 marks*)

Free radicals

- Free radicals are very reactive species.
- They are formed when a covalent bond breaks so that one electron is transferred to each of the atoms.
- This is called homolytic fission and forms two free radicals.

$$A-B \rightarrow A\bullet + B\bullet$$

Notice that the dot, which represents the unpaired electron, is written next to the atom that has the unpaired electron.

The free radical substitution of methane

Methane reacts with chlorine by a free radical substitution reaction. You can sum up this reaction using the equation:

$$CH_4 + Cl_2 \rightarrow CH_3Cl + HCl$$

Initiation

- UV light provides the energy required for the homolytic fission of the chlorine molecule.

$$Cl_2 \rightarrow 2Cl\bullet$$

Propagation

- Next the chlorine free radical reacts with methane to form hydrogen chloride and a methyl free radical.

$$Cl\bullet + CH_4 \rightarrow HCl + \bullet CH_3$$

- Then the methyl free radical reacts to form chloromethane and a chlorine free radical.

$$\bullet CH_3 + Cl_2 \rightarrow CH_3Cl + Cl\bullet$$

Notice that as the free radical reacts it forms a new molecule and another free radical.

Termination

In the termination step two free radicals react together to form a new molecule.

$$Cl\bullet + Cl\bullet \rightarrow Cl_2$$
$$\bullet CH_3 + Cl\bullet \rightarrow CH_3Cl$$
$$\bullet CH_3 + \bullet CH_3 \rightarrow C_2H_6$$

Notice that the three steps in the free radical substitution of methane are initiation, propagation, and termination.

- In the initiation step free radicals are made.
- In the propagation steps free radicals react with molecules to form new molecules and free radicals.
- In the termination steps two free radicals join together. A variety of new molecules are formed.

1 An unbranched saturated hydrocarbon A has the molecular formula C_5H_{12}.

   a Define the term *saturated hydrocarbon*.   (*2 marks*)

   b Name hydrocarbon A.   (*1 mark*)

2 Methane and coal are fuels.

   a Define the term *fuel*.   (*1 mark*)

   b Give the equation for the complete combustion of methane.   (*2 marks*)

   c Explain how acid rain is formed when fossil fuels, like coal, are burnt.   (*2 marks*)

3 The catalytic cracking of hydrocarbon molecules takes place at a lower temperature and pressure than is used in thermal cracking.

   a Name the catalyst used in the catalytic cracking of hydrocarbons.   (*1 mark*)

   b Explain why this catalyst has a honeycomb structure.   (*1 mark*)

4 State and explain the trend in boiling point down the alkane homologous series.   (*3 marks*)

5 Propane is a useful fuel.

   a Give the equation for the complete combustion of propane.   (*2 marks*)

   b Other than carbon dioxide and water vapour name one other product made during the complete combustion of propene.   (*1 mark*)

   c Why does incomplete combustion occur?   (*1 mark*)

6 Methane reacts with chlorine to produce chloromethane in a free radical substitution reaction.

   a Give the overall equation for the reaction between chlorine and methane.   (*1 mark*)

   b Define the term *radical*.   (*1 mark*)

   The equation for the initial step is shown below.

   $Cl_2 \rightarrow 2Cl\bullet$

   c What conditions are required in the initial step?   (*1 mark*)

   The propagation step of the free radical substitution reaction takes place in two stages. In the first stage

   $Cl\cdot + CH_4 \rightarrow HCl + CH_3\bullet$

   d Give the equation for the second step in the propagation step.   (*1 mark*)

   e The termination step can produce ethane. Give the equation for the termination step that produces ethane.   (*1 mark*)

7 Which of the following compounds is a member of the alkane homologous series?   (*1 mark*)

   a C_7H_{14}

   b C_7H_{16}

   c C_2H_4

   d CH_2

8 How many atoms are present in one molecule of butane?   (*1 mark*)

   a 14

   b 12

   c 9

   d 16

13.1 Halogenoalkanes – introduction

Specification reference: 3.3.3

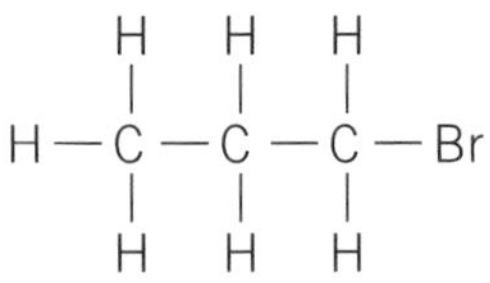

▲ **Figure 1** *1-bromopropane*

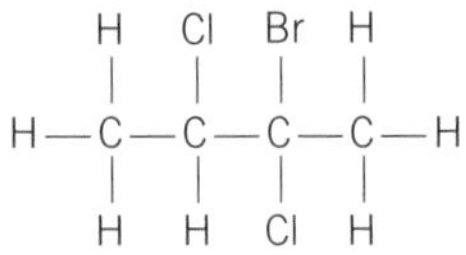

▲ **Figure 2** *2-bromo-2, 3-dichloro butane*

Halogenoalkanes

The halogenoalkanes, $C_nH_{2n+1}X$, are alkanes in which one hydrogen atom has been substituted by a halogen atom, X, where X can be fluorine, chlorine, bromine, or iodine. Their names may include a number to indicate where the halogen atom is bonded onto the carbon chain.

More complicated halogenoalkanes

The prefix di, tri, tetra is sometimes included in names to show the number of each type of halogen atom present. Some compounds contain two different halogen atoms. These halogenoalkanes are named by placing the halogens in alphabetical order.

Why are halogenoalkanes more reactive than alkanes?

Polar bonds

▼ **Table 1** *Table of electronegativity values*

Element	Electronegativity
C	2.5
F	4.0
Cl	3.5
Br	2.8
I	2.6

Halogenoalkanes contain a carbon–halogen bond. Halogen atoms are more electronegative than carbon and this results in a polar C–X bond being formed. The greatest difference in electronegativity is in the C–F bond, so the C–F bond is the most polar C–X bond. Bond polarity decreases down the group. These polar bonds mean halogenoalkanes can be attacked by nucleophiles which makes halogenoalkanes more reactive than alkanes.

Strength of the C–X bond

The strength of the C–X bond also influences the reactivity of halogenoalkanes. The strength of the C–X bond decreases down the group. Smaller atoms attract the shared pair of electrons in the C–X bond more strongly so have the greatest bond enthalpy. Down the group the halogen atoms become larger so their attraction for the shared pair of electrons in the C–X bond decreases and the bond becomes weaker.

▼ **Table 2** *Table of bond enthalpy values*

Bond	Bond enthalpy / $kJ\,mol^{-1}$
C–F	467
C–H (for comparison)	413
C–Cl	346
C–Br	290
C–I	228

Table 2 shows that C–Cl, C–Br, and C–I bonds found in chloroalkanes, bromoalkanes, and iodoalkanes are all weaker than the C–H bonds found in alkanes. Many halogenoalkanes are more reactive than alkanes.

Experiments show that the halogenoalkanes become more reactive down the group. This proves that bond enthalpy is a more important factor than bond polarity in predicting the reactivity of halogenoalkanes.

Summary questions

1 Name these halogenoalkanes:

a

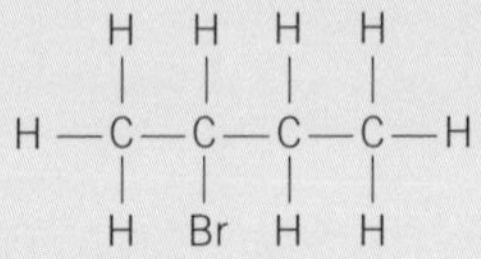

b

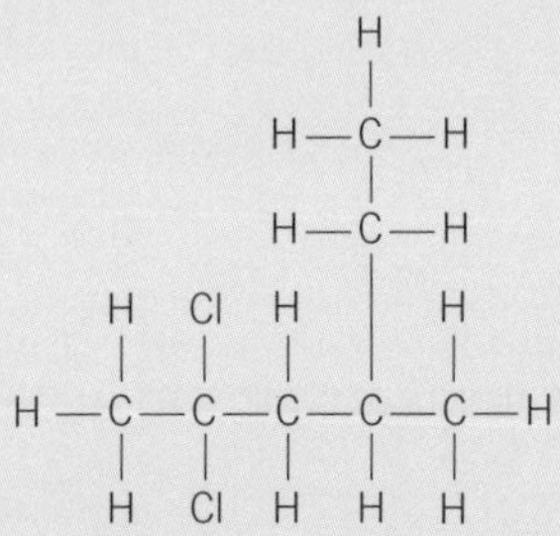

(2 marks)

2 State and explain the trend in the bond polarity of the C–X bond down the halogen group. *(2 marks)*

3 State and explain the trend in the bond enthalpy of the C–X bond down the halogen group. *(2 marks)*

13.2 Nucleophilic substitution in halogenoalkanes

Specification reference: 3.3.3

The nucleophilic substitution of halogenoalkanes

In the nucleophilic substitution of a halogenoalkane a nucleophile is added to a molecule and a halide ion is lost.

Hydroxide ions

Hydroxide, OH^-, ions are nucleophiles which have a lone pair of electrons on the oxygen atom which they can donate to form a new covalent bond. Halogenoalkanes undergo nucleophilic substitution reactions with hydroxide ions (from sodium hydroxide or potassium hydroxide) to form alcohols. Halogenoalkanes do not dissolve in water so ethanol is used as a solvent to allow both reactants to mix and react together. This reaction can also be described as hydrolysis.

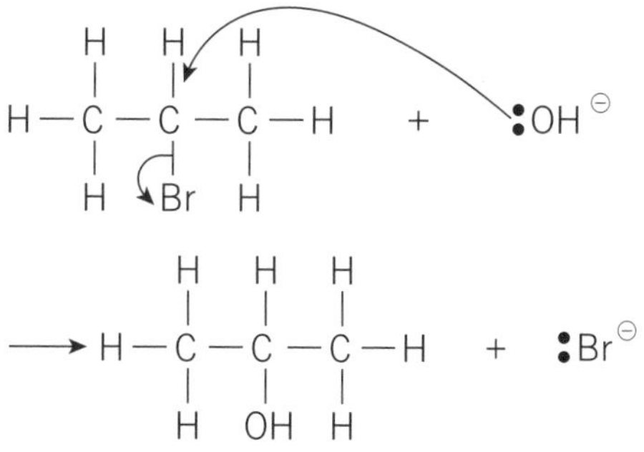

▲ **Figure 1** *Nucleophilic substitution by hydroxide ions*

Cyanide ions

Cyanide, ^-CN, ions are nucleophiles that have a lone pair of electrons on the nitrogen atom which they can donate to form a new covalent bond. Halogenoalkanes undergo nucleophilic substitution reactions with cyanide, ^-CN, ions to form nitriles. The halogenoalkane is warmed with potassium cyanide. This reaction is a useful way of increasing the carbon chain length.

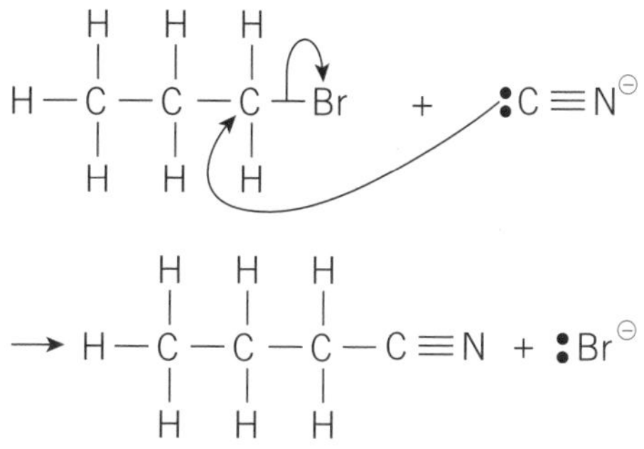

▲ **Figure 2** *Nucleophilic substitution by cyanide ions*

Ammonia molecules

Ammonia, NH_3, molecules are nucleophiles that have a lone pair of electrons on the nitrogen atom which they can donate to form a new covalent bond.

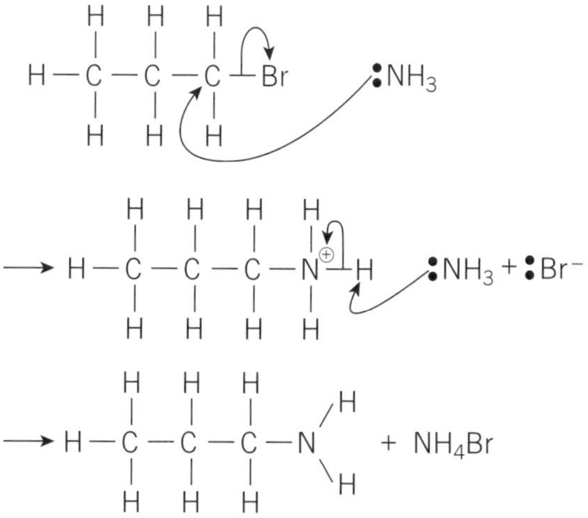

▲ **Figure 3** *Nucleophilic substitution by ammonia molecules*

Halogenalkanes undergo nucleophilic substitution reactions with ammonia, NH_3, molecules to form amines. The halogenoalkanes react with concentrated ammonia solution in ethanol at high pressure.

Using nucleophilic substitution reactions

The nucleophilic substitution reactions of halogenoalkanes are used to introduce new functional groups to organic molecules. In particular using cyanide, ^-CN, ions allows us to increase the carbon chain length.

Key terms

Reaction mechanism: The route or journey from the reactants to the products through a series of theoretical steps.

Substitution: When a substance swaps one atom or group for another atom or group.

Hydrolysis: When a substance is split up using water molecules or hydroxide ions.

Nucleophiles: Nucleophiles such as ^-OH, ^-CN, and NH_3 have a lone pair of electrons which they can donate to another molecule to form a new covalent bond.

Summary questions

1. What is a nucleophile? *(1 mark)*
2. Why is ethanol used in the reaction between 1-bromopropane and sodium hydroxide? *(1 mark)*
3. Give the mechanism for the nucleophilic substitution reaction of 2-chloropropane with cyanide. *(4 marks)*

13.3 Elimination reactions in halogenoalkanes

Specification reference: 3.3.3

Key terms

Primary halogenoalkanes: The halogen is bonded to a carbon which is bonded to just one other carbon atom.

▲ **Figure 1** *1-bromopropane*

Secondary halogenoalkanes: The halogen is bonded to a carbon which is bonded to two other carbon atoms.

▲ **Figure 2** *2-bromopropane*

Tertiary halogenoalkanes: The halogen is bonded to a carbon which is bonded to three other carbon atoms.

▲ **Figure 3** *2-bromo-2-methylpropane*

Elimination reactions

In elimination reactions a small molecule is lost from an organic molecule. When halogenoalkanes undergo elimination reactions a hydrogen halide is lost and an alkene is formed.

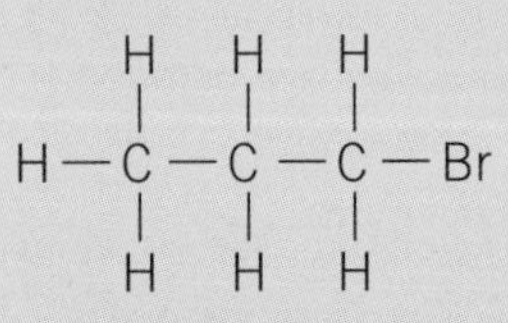

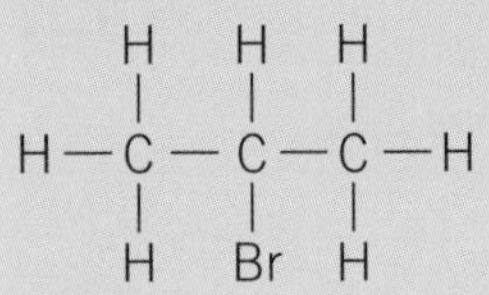

The role of hydroxide ions

Hydroxide, OH^-, ions act as nucleophiles in nucleophilic substitution reactions. However under different conditions the hydroxide ions can act as bases. This happens in elimination reaction.

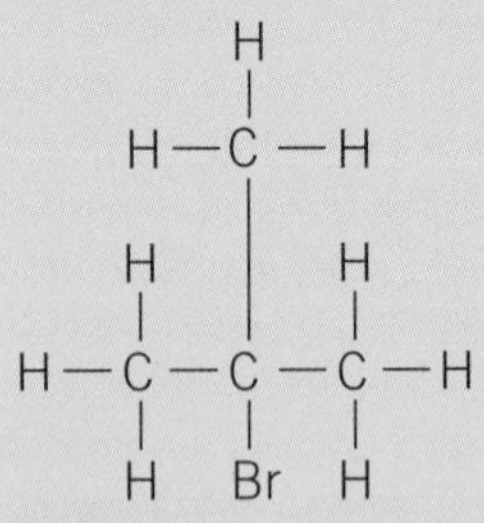

The importance of conditions used

The reaction between halogenoalkanes and hydroxide ions can produce different products depending on the conditions chosen. If the reactants are dissolved in water and the reaction is carried out at room temperature nucleophilic substation reactions are favoured. If the reactants are dissolved in alcohol (ethanol) and the reaction is carried out at high temperature elimination is favoured.

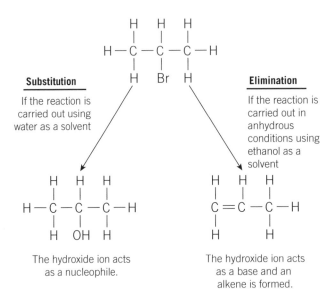

▲ **Figure 4** *The importance of conditions*

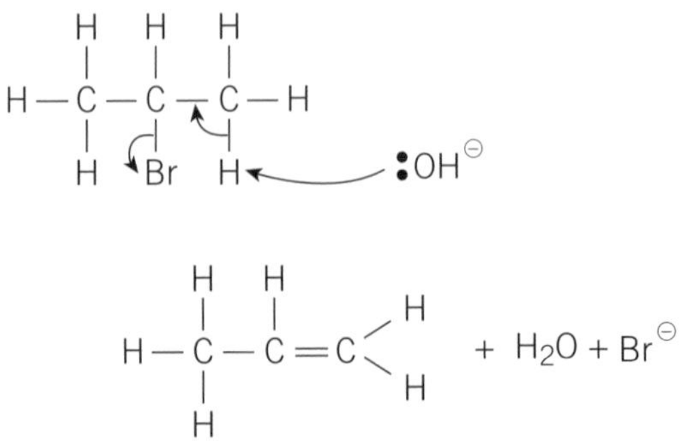

▲ **Figure 5** *Mechanism for the elimination reaction*

Types of halogenoalkane

Primary halogenoalkanes favour nucleophilic substitution reactions while tertiary halogenoalkanes favour elimination reactions. Secondary halogenoalkanes react readily by either mechanism.

Making isomeric alkenes

Elimination reactions can produce a number of different products which are structural and or stereoisomers of each. The reaction below produces three isomers.

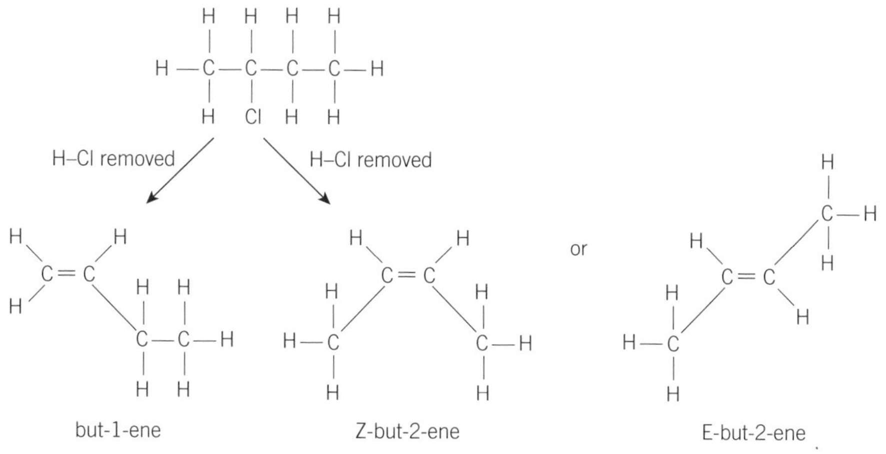

▲ Figure 6

Key term

Elimination reaction: In elimination reactions a small molecule, such as water, is lost from an organic molecule.

Synoptic link

Decolourising bromine water is a test for an alkene. The bromine adds on across the double bond. See Topic 14.2, Reactions of alkenes.

Key terms

Base: Bases are proton acceptors.

Structural isomer: Structural isomers have the same molecular formula but a different structural formula.

Stereoisomers: Stereoisomers have the same structural formula but a different arrangement of atoms in space.

Summary questions

1 Define the term *elimination reaction*. (*1 mark*)

2 Name the organic product of the reaction between 2-bromopropane and hydroxide ions in aqueous conditions. (*1 mark*)

3 Give the displayed formula and name the organic products of the reaction between 2-chloropentane and ethanolic potassium hydroxide. (*3 marks*)

1 Give the skeletal formula and names of the four organic molecules with the molecular formula $C_3H_6Cl_2$. *(4 marks)*

2 A student has a sample of a halogenoalkane which they believe to be a bromoalkane. Outline a method that the student can carry out to confirm that the sample is a bromoalkane. Include any observations that the student should make. *(4 marks)*

3 Consider the two halogenoalkanes below:

CH_3CH_2Cl

$CH_3(CH_2)_3Cl$

   a Name each of these halogenoalkanes. *(2 marks)*

   b These two halogenoalkanes have different boiling points. State and explain which of these halogenoalkanes has the higher boiling point. *(3 marks)*

4 Bromomethane reacts with an aqueous solution of sodium hydroxide. One of the products is an alcohol. The sodium hydroxide solution contains hydroxide ions which act as nucleophiles.

   a Define the term *nucleophile*. *(1 mark)*

   b Complete the diagram to show mechanism for the reaction between bromomethane and hydroxide ions. *(4 marks)*

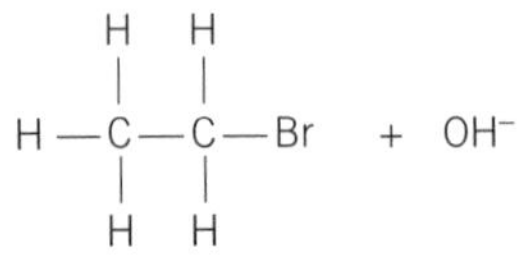

   c Name the reaction mechanism for the reaction between bromomethane and aqueous hydroxide ions. *(2 marks)*

5 Sodium hydroxide is dissolved in ethanol and mixed with 1-chlorobutane. The reaction mixture is then heated. One of the products is an alkene.

   a Show the reaction mechanism for the reaction that takes place. *(4 marks)*

   b Name the organic product of the reaction. *(1 mark)*

   c Name the reaction mechanism for the reaction between bromomethane and ethanolic hydroxide ions. *(4 marks)*

6 Which C–X bond is the most polar?

   A C–F

   B C–Cl

   C C–Br

   D C–I *(1 mark)*

7 Which C–X bond has the greatest bond enthalpy?

   A C–F

   B C–Cl

   C C–Br

   D C–I *(1 mark)*

14.1 Alkenes

Specification reference: 3.3.4

Introducing alkenes

Alkenes are unsaturated hydrocarbons with the general formula C_nH_{2n}.

Alkenes have a C=C double bond that consists of a sigma bond and a pi bond. The pi bond forms above and below the axis of the carbon atoms.

Ethene is a planar (flat) molecule because of the C=C arrangement in alkene molecules.

The pi bond has a high electron density (a high chance of the electron being present). Electrophiles can attack this high electron density.

As a result alkenes are more reactive than alkanes.

Key terms

Unsaturated: An unsaturated hydrocarbon contains at least one carbon double bond.

Electrophile: A species that can accept a lone pair of electrons to form a new covalent bond.

E–Z stereoisomers

Rotation of the C=C bond would require the pi bond to be broken.

This requires a lot of energy so there is restricted rotation about the C=C double bond.

As a result but-2-ene exists as two stereoisomers.

Notice that in the *E* form the methyl groups are arranged on opposite sides of the C=C bond. The *E* comes from the German word *entgegen* which means opposite.

In the *Z* form the methyl groups are arranged on the same side of the C=C bond. The *Z* comes from the German word *zusammen* which means together.

But-2-ene has stereoisomers because it has two different groups attached to each of the double bonded carbon atoms.

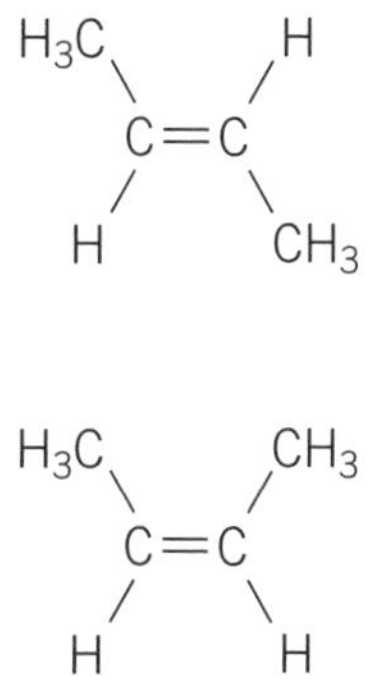

E-but-2-ene

Z-but-2-ene

▲ **Figure 1** *E–Z stereoisomers*

Reactions of alkenes

The bond enthalpy of a C–C bond is 347 kJ mol^{-1} while the bond enthalpy of the C=C bond is 612 kJ mol^{-1}. However, alkenes are much more reactive than alkanes due to the electron-rich C=C bond in the alkene molecules. In addition reactions two species join together. An electrophile is a species that can accept a pair of electrons to form a new covalent bond. Electrophiles can attack the high electron density of the pi bond in alkenes. As a result alkenes undergo electrophilic addition reactions.

Reaction with bromine

Alkenes react with bromine to form dibromoalkanes. For example:

$$C_2H_4 + Br_2 \rightarrow C_2H_4Br_2$$

This is an electrophilic addition reaction.

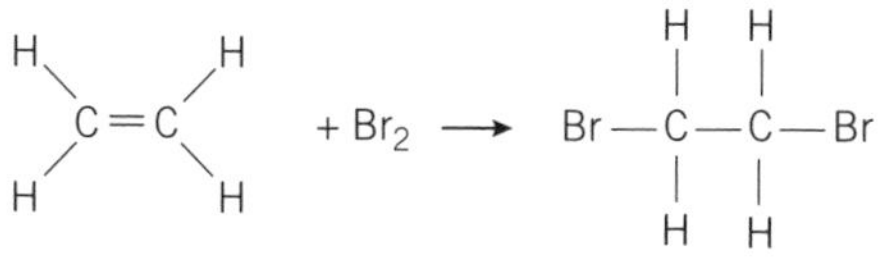

▲ **Figure 2** *The reaction mechanism for the electrophilic addition of bromine to ethene*

Summary questions

1. Define the term *electrophile*. *(1 mark)*
2. Name the type of reaction that takes place between chlorine and ethene. *(1 mark)*
3. Draw the displayed formula and name the three alkene molecules which are isomers of each other and have the molecular formula C_4H_8. *(3 marks)*

14.2 Reactions of alkenes

Specification reference: 3.3.4

Electrophilic addition reactions

The C=C bond in alkene molecules is an electron-rich area that can be attacked by electrophiles. Electrophiles are species than can accept a pair of electrons to form a new covalent bond.

Alkene molecules often undergo electrophilic substitution reactions.

Reaction with hydrogen bromide

Alkenes react with hydrogen bromide to form bromoalkanes. For example:

$$C_2H_4 + HBr \rightarrow C_2H_5Br$$

This is an electrophilic addition reaction.

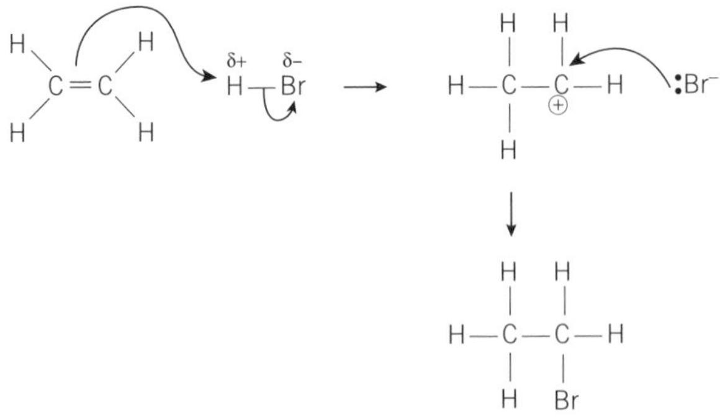

▲ **Figure 1** *The reaction mechanism for the electrophilic addition of hydrogen bromide to ethene*

Revision tip

A useful guide for predicting the major product of an addition reaction with an unsymmetrical alkene is to use Markovnikov's rule.

Markovnikov's rule states that the hydrogen goes to the carbon atom which has the most hydrogens already.

Unsymmetrical alkenes

Ethene is a symmetrical alkene. The groups attached to the C=C bond are the same.

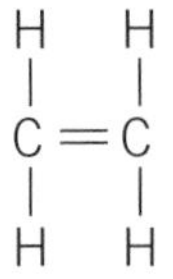

▲ **Figure 2** *Ethene*

When symmetrical alkenes such as ethene take part in addition reactions with electrophiles such as hydrogen bromide it does not matter how the hydrogen bromide adds across the carbon double bond because the product is always bromoethane.

However, this is not true for all alkene molecules.

Propene is an unsymmetrical alkene.

Notice that the groups attached to the carbon atoms in the C=C bond are different. This is significant when propene takes part in an addition reaction with an electrophile such as HBr. The HBr can add across the bond in two different ways, to produce 2-bromopropane or 1-bromopropane.

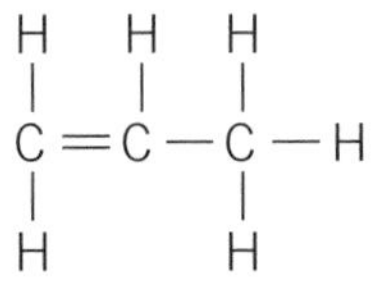

▲ **Figure 3** *Propene*

Summary questions

1 Name the reaction mechanism favoured by alkenes. *(1 mark)*

2 Complete the reaction mechanism below to show the reaction between chlorine and ethene.

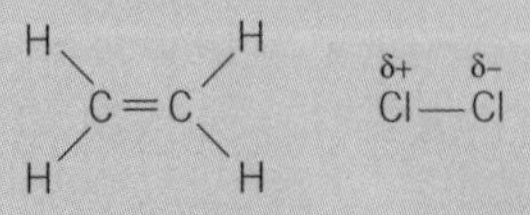

(4 marks)

3 Draw the displayed formula of the two products of the electrophilic addition of hydrogen chloride to propene. *(2 marks)*

14.3 Addition polymers

Specification reference: 3.3.4

Polymers

Lots of alkene molecules can be joined together to form addition polymers.

- The alkene molecules are called monomers.
- This is an addition polymerisation reaction.

Many ethene molecules can join together to form the addition polymer poly(ethene).

Notice that poly(alkenes) are saturated – they do not have C=C double bonds.

Many propene molecules can join together to form poly(propene).

The section drawn in brackets is called a repeating unit. The repeating unit is repeated thousands of times in each polymer molecule. The lines representing the covalent bonds between repeating units cross through the brackets around the repeating units.

Identifying the monomer

The monomer used to produce an addition polymer can be identified from the repeat unit of the polymer.

First the trailing bonds (bonds through the brackets) are removed.

Then replace the single covalent bond between the carbon atoms with a double covalent bond.

Why are poly(alkenes) unreactive?

- Alkenes have a sigma and a pi bond.
- The high electron density and easy accessibility of the pi bond mean that alkenes can be attacked by electrophiles.
- Poly(alkenes) are saturated hydrocarbons.
  - As a result they are much less reactive than alkenes.

Key term

Addition polymer: An addition polymer is formed when lots of small monomers are chemically joined together to form a larger polymer molecule.

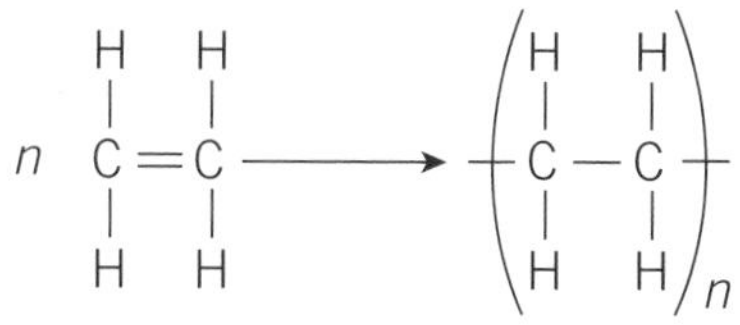

▲ **Figure 1** *The formation of poly(ethene)*

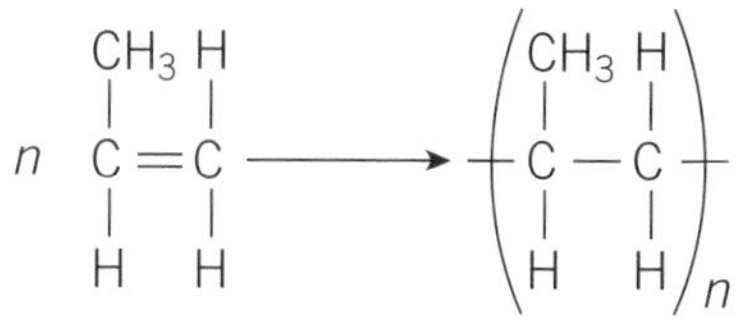

▲ **Figure 2** *The formation of poly(propene)*

Common misconception

Both sulfur dioxide and nitrogen oxides cause acid rain.

Summary questions

1 Name the polymer produced from the addition polymerisation of the following monomers:
ethene   chloroethene   propene *(3 marks)*

2 Consider the monomer below.

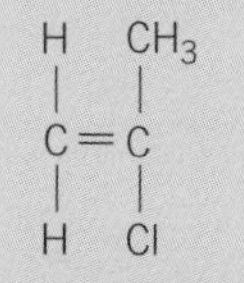

Draw the repeat unit of the addition polymer produced from this monomer. *(1 mark)*

3 Consider the repeat unit of the addition polymer shown below. Deduce the monomer used to make this polymer. *(1 mark)*

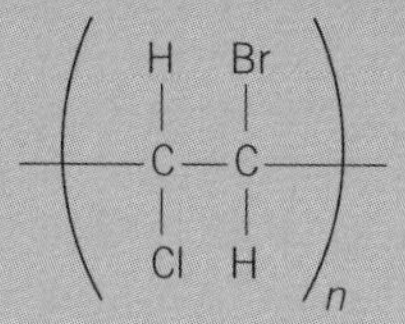

Chapter 14 Practice questions

1 Ethene reacts with hydrogen chloride to form a new organic compound.

    a Name the new organic compound formed. *(1 mark)*

    b Complete the reaction mechanism for the reaction between ethene and hydrogen chloride. *(5 marks)*

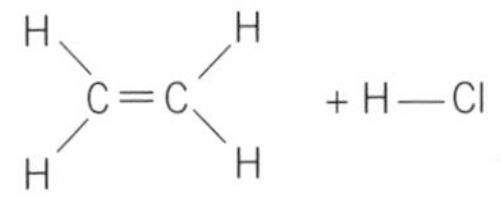

2 A styrene molecule has the displayed formula below.

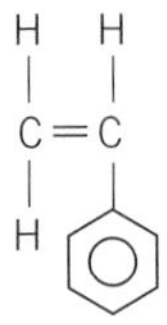

    a Styrene molecules can be used to make polystyrene.

       Draw the repeat unit of polystyrene. *(1 mark)*

    b Name the type of reaction that produced polystyrene. *(1 mark)*

3 Propene is a fuel.

    a Define the term *fuel*. *(1 mark)*

    b Give the equation for the complete combustion of propene. *(2 marks)*

4 Alkenes are unsaturated hydrocarbons.

    a Define the term *unsaturated hydrocarbon*. *(1 mark)*

The names and molecular formula of the first five members of the alkene homologous series are shown in the table below.

Name	Molecular formula
ethene	C_2H_4
propene	C_3H_6
butene	C_4H_8
pentene	C_5H_{10}
hexene	C_6H_{12}

    b What is the general formula of alkenes? *(1 mark)*

    c Predict the molecular formula of the seventh member of the alkene homologous series. *(1 mark)*

5 A hydrocarbon is analysed and contains 85.7% carbon by weight. The hydrocarbon has a relative molecular mass of 112.0.

    a Calculate the empirical formula of the hydrocarbon. *(2 marks)*

    b Deduce the molecular formula of the hydrocarbon. *(1 mark)*

6 Give the molecular formula of the monomer that is used to produce polypropene.

    A C_3H_8

    B C_3H_6

    C C_2H_4

    D C_2H_6 *(1 mark)*

7 The test for a carbon double bond is:

    A It turns limewater cloudy.

    B A white precipitate forms when silver nitrate is added.

    C A cream precipitate forms when silver nitrate is added.

    D It decolourises bromine water. *(1 mark)*

15.1 Alcohols – an introduction

Specification reference: 3.3.5

Alcohols

Ethanol is an alcohol.

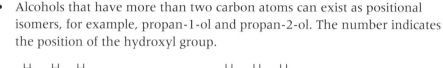
▲ **Figure 1** *Ethanol*

- Alcohols have the general formula $C_nH_{2n+1}OH$.
- Alcohols contain the –OH or hydroxyl group.
- Alcohols that have more than two carbon atoms can exist as positional isomers, for example, propan-1-ol and propan-2-ol. The number indicates the position of the hydroxyl group.

propan-1-ol

propan-2-ol

▲ **Figure 2** *Positional isomers*

Types of alcohol

We can classify alcohols as primary, secondary, or tertiary depending on the number of carbon atoms attached to the carbon atom bonded to the hydroxyl, –OH, group.

Primary alcohols

In primary alcohols the carbon atom attached to the hydroxyl group is attached to one other carbon atom.

▲ **Figure 3** *A primary alcohol*

Secondary alcohols

In secondary alcohols the carbon atom attached to the hydroxyl group is attached to two other carbon atoms.

▲ **Figure 4** *A secondary alcohol*

Tertiary alcohols

In tertiary alcohols the carbon atom attached to the hydroxyl group is attached to three other carbon atoms.

▲ **Figure 5** *A tertiary alcohol*

Physical properties of alcohols

Alcohols contain a hydroxyl, –OH, group which means that hydrogen bonds form between alcohol molecules. These strong intermolecular forces of attraction mean that alcohols have relatively high melting points and boiling points compared with alkanes of a similar molecular mass.

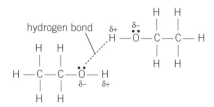

◀ **Figure 7** *Hydrogen bonding*

Alcohols can also form hydrogen bonds to water molecules. This means that short carbon chain length alcohols are soluble in water.

Worked example

Q Draw the structural formula of pentan-2-ol.

A

pentan-2-ol

Worked example

Q Draw the structural formula of the tertiary alcohol with the formula $C_4H_{10}O$.

A

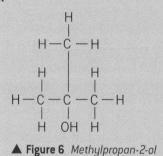

▲ **Figure 6** *Methylpropan-2-ol*

Summary questions

1 What is the general formula of an alcohol?
   (1 mark)

2 What type of alcohol is propan-2-ol? *(1 mark)*

3 Why is methanol soluble in water? Your answer should include a diagram. *(2 marks)*

15.2 Ethanol production

Specification reference: 3.3.5

Producing ethanol

Ethanol can be produced by:

- hydration of ethene
- fermentation.

Hydration of ethene

Ethanol can be made industrially by the hydration of ethene

$$C_2H_4 + H_2O \rightarrow C_2H_5OH$$

This is an addition reaction. Addition reactions have an atom economy of 100%. This is good for sustainable development. All the reactant atoms are made into useful products.

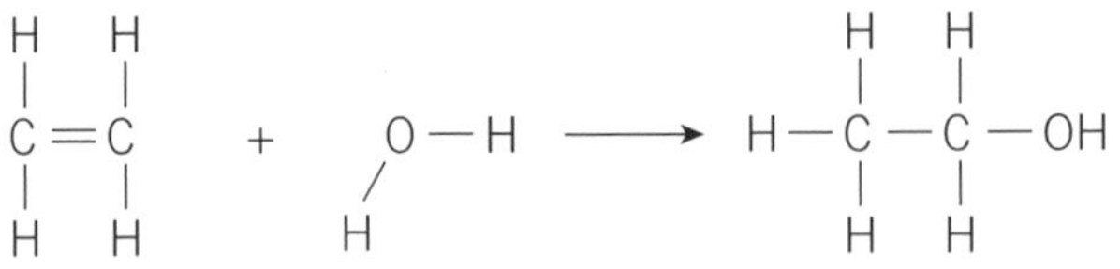

▲ **Figure 1** *Producing ethanol*

Fermentation 🔬

Ethanol can also be made industrially by the fermentation of glucose from plants.

$$glucose \rightarrow ethanol + carbon\ dioxide$$
$$C_6H_{12}O_6 \rightarrow 2CH_3CH_2OH + 2CO_2$$

Fermentation versus hydration of ethane Biofuels

Fermentation	Hydration of ethene
Slower rate of reaction.	Faster rate of reaction.
Yield of ethanol of around 15%.	Yield of ethanol of around 95%.
Batch process. It is more labour intensive, so labour costs are higher but set up costs are lower.	Continuous process, so labour costs are less but set-up costs are higher.
Atom economy of 51.1%.	An addition reaction with an atom economy of 100%.
Uses lower temperatures and pressures so uses lower amounts of energy.	Uses higher temperatures and pressures so uses higher amounts of energy.
Renewable.	Non-renewable.
Distillation is used to increase the concentration of alcohol. The alcohol that is produced has a lower purity so more steps are required for purification.	Distillation is used to increase the concentration of alcohol. The alcohol produced is already purer so purification is easier.

Summary questions

1 Give two ways that ethanol can be made.
(1 mark)

2 Which method of making ethanol produces an alcohol that is a renewable? Explain your answer.
(1 mark)

3 The equation for the hydration of ethene and the fermentation of sugar to make ethanol are shown below.

   a Hydration of ethene
   $C_2H_4 + H_2O \rightarrow C_2H_5OH$

   b Fermentation
   $C_6H_{12}O_6 \rightarrow 2CH_3CH_2OH + 2CO_2$

   Calculate the atom economy of both these methods of making ethanol. Give your answers to two significant figures.
(2 marks)

15.3 The reactions of alcohols

Specification reference: 3.3.5

Primary and secondary alcohols ⚗

Primary and secondary alcohols are oxidised by oxidising agents such as aqueous acidified potassium dichromate(VI), $K_2Cr_2O_7$.

Primary alcohols

Primary alcohols are first oxidised to aldehydes. The aldehyde can be distilled off as it is made to stop any further oxidation.

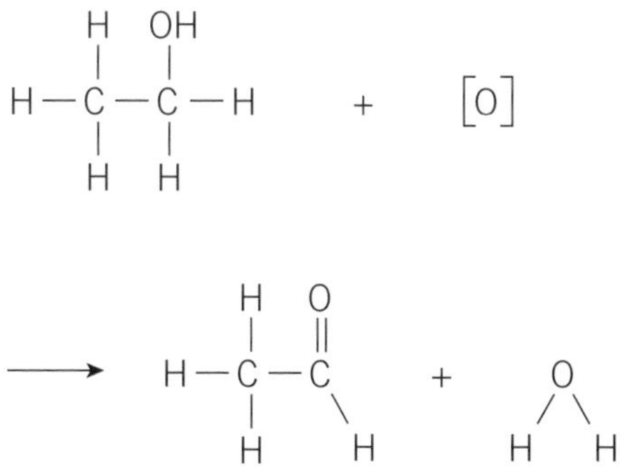

◀ **Figure 1** *Production of aldehydes*

$$CH_3CH_2OH + [O] \rightarrow CH_3CHO + H_2O$$

Notice how the oxidising agent is written as [O] but it is important that the equation still balances.

As the primary alcohols is oxidised the chromium(VI) ion is reduce to chromium(III) and the solution changes colour from orange to green.

If the oxidising agents are in excess and the reaction mixture is heated under reflux then the aldehyde is oxidised to a carboxylic acid.

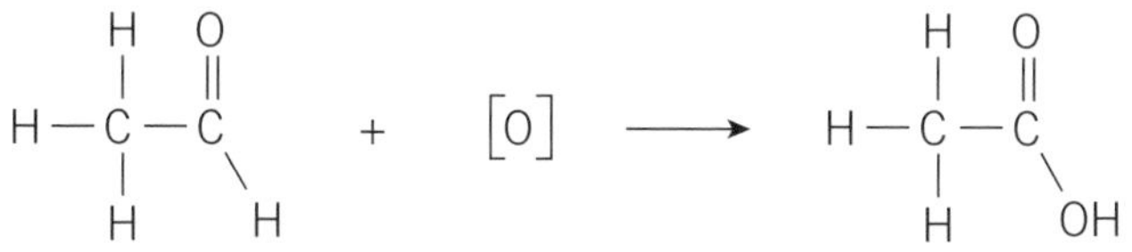

▲ **Figure 2** *Production of carboxylic acids*

$$CH_3CHO + [O] \rightarrow CH_3COOH$$

As the aldehyde is oxidised the chromium(VI) ion is reduced to chromium(III) and the solution changes colour from orange to green.

Secondary alcohols

Secondary alcohols are oxidised to ketones but they cannot be oxidised any further.

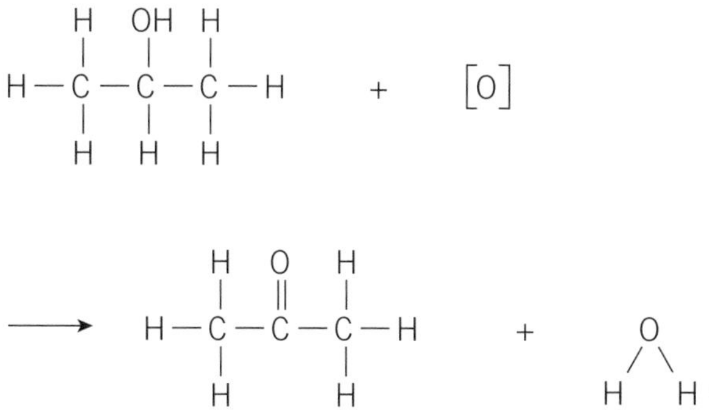

◀ **Figure 3** *Production of ketones*

$$CH_3CH(OH)CH_3 + [O] \rightarrow CH_3COCH_3 + H_2O$$

Tertiary alcohols

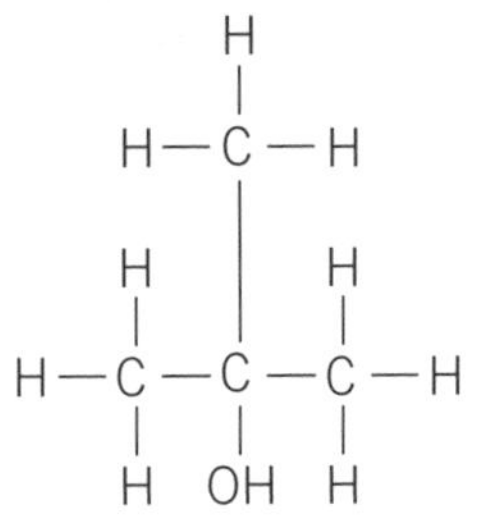

▲ **Figure 4** *Mathylpropan-2-ol*

Tertiary alcohols such as methylpropan-2-ol have a hydroxyl group attached to a carbon atom that is attached to three other carbon atoms.

- Tertiary alcohols are not oxidised by oxidising agents such as aqueous acidified potassium dichromate(VI), $K_2Cr_2O_7$. As a result if acidified potassium dichromate(VI) solution is added to a tertiary alcohol no oxidation reaction takes place so the orange solution stays orange.

Distinguishing between aldehydes and ketones

Aldehydes and ketones both contain carbonyl, C=O, groups.

Aldehydes

In aldehydes the carbonyl is at the end of the carbon chain.

Ketones

In ketones the carbonyl is not at the end of the carbon chain.

Tollens' reagent

Tollens' reagent, ammoniacal silver nitrate, is a mild oxidising agent that can be used to distinguish aldehydes from ketones. When Tollens' reagent is added to aldehydes they are oxidised to carboxylic acids and the silver Ag^+ ions in the Tollens' reagent are reduced to silver atoms, Ag, and a silver mirror forms.

Ketones cannot be oxidised so when Tollens' reagent is added to ketones there is no reaction.

Fehling's solution

Fehling's solution is another mild oxidising agent that can be used to distinguish between aldehydres and ketones. It is a blue solution that contains a complex of Cu^{2+} ions.

When Fehling's solution is heated with an aldehyde the aldehyde is oxidised to a carboxylic acid and the Cu^{2+} ions are reduced to Cu^+ ions. A brick-red precipitate of Cu_2O forms.

Ketones cannot be oxidised, so no reaction happens and the blue Fehling's solution does not change colour.

Dehydrating alcohols

Dehydrating agents such as concentrated sulfuric acid or phosphoric acid can be used to remove H_2O from alcohols to form alkenes.

Example

propan-1-ol → propene + water

$CH_3CH_2CH_2OH \rightarrow CH_3CHCH_2 + H_2O$

Notice that as water is lost from the alcohol during the reaction. This may also be classified as an elimination reaction.

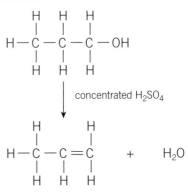

▲ **Figure 5** *Production of alkenes*

Key term

Elimination reaction: In elimination reactions a small molecule is lost.

Summary questions

1 What is formed when a secondary alcohol is oxidised? *(1 mark)*

2 Give two reagents that could be used to differentiate between an aldehyde and a ketone. *(2 marks)*

3 Suggest a dehydrating agent that could be used to form an alkene from an alcohol. *(1 mark)*

4 Propan-1-ol is heated with an excess of aqueous acidified potassium dichromate(VI), $K_2Cr_2O_7$. The reaction mixture is heated under reflux.

a Name the organic product formed.

b Give the equation for the reaction. *(2 marks)*

Chapter 15 Practice questions

1 A sample of an unbranched organic compound was analysed and found to contain only carbon, hydrogen, and oxygen. The sample contained 68.2% carbon and 18.2% oxygen by mass. The relative molecular mass of the organic compound was found to be 88.0.

   a   Calculate the empirical formula of the compound.   (*2 marks*)

   b   Deduce the molecular formula of the compound.   (*1 mark*)

2 Propan-1-ol can be dehydrated to produce useful new products.

   a   Define the term *dehydration*.   (*1 mark*)

   b   Identify a suitable catalyst for the dehydration of propan-1-ol.   (*1 mark*)

   c   Name the organic product produced by the dehydration of propan-1-ol.   (*1 mark*)

3 Butan-1-ol and 2-methylpropan-2-ol are structural isomers.

   a   What is the molecular formula shared by both compounds?   (*1 mark*)

   b   Define the term *structural isomers*.   (*1 mark*)

   c   Explain how a student could use a solution of acidified potassium dichromate(VI) to differentiate between these two compounds.   (*4 marks*)

4 Primary alcohols can be oxidised by acidified potassium dichromate(VI) to produce useful new organic compounds. The amount of oxidation can be controlled by controlling the conditions used during oxidation. Complete the equations below to show the products of oxidation in each reaction.   (*3 marks*)

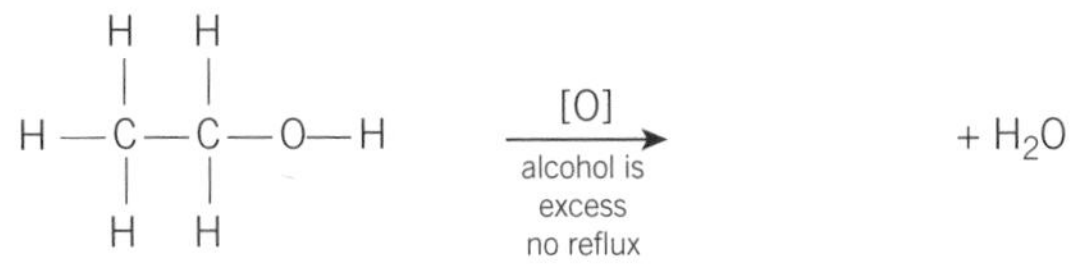

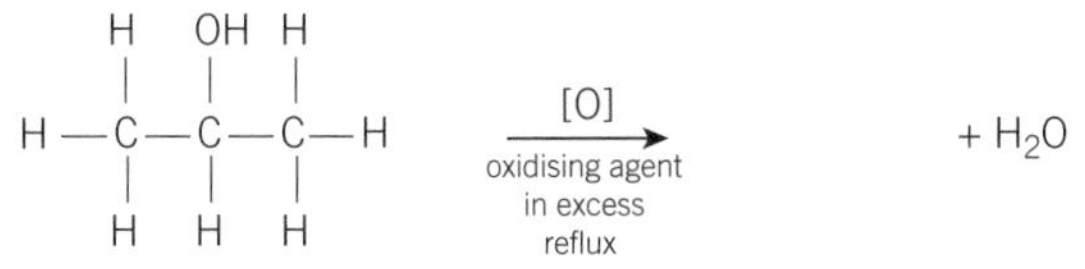

5 Hexanal can be oxidised to a carboxylic acid.

   a   Name the carboxylic acid formed in this reaction.   (*1 mark*)

   b   Hexanal can be produced by the oxidation of an alcohol.

       i   Identify this alcohol.   (*1 mark*)

       ii   State the class to which the alcohol belongs.   (*1 mark*)

6 Name the organic product made when butan-2-ol is heated under reflux with acidified potassium dichromate(VI).

   A   butanal

   B   butan-2-one

   C   butan-3-one

   D   butanoic acid   (*1 mark*)

16.1 Test-tube reactions

Revision tip

Alkene molecules contain the C=C functional group. Halogenoalkane molecules contain a C-X bond where X is a halogen atom. Alcohols contain the hydroxyl, OH group.

Summary questions

1 How could you confirm that a hydrocarbon was an alkane? (*1 mark*)

2 A student adds an aqueous solution of sodium hydrogencarbonate to a sample of an organic compound. The student observed effervescence. What can the student deduce about the organic compound? (*1 mark*)

3 A sample of an alcohol was heated under reflux with an excess of acidified potassium dichromate(VI) but the reaction mixture stayed orange. What type of alcohol was present? (*1 mark*)

4 A student warms a sample of a carbonyl compound with Tollen's reagent. The student observed that a silver mirror was formed. What can the student deduce about the carbonyl compound being investigated? (*1 mark*)

5 A student has a sample of a halogenoalkane. Outline the steps the student must carry out to confirm that the halogenoalkane is a chloroalkane. (*2 marks*)

Identifying organic compounds

The groups in organic compounds can be identified by some straightforward tests.

Functional group	Test	Result	Notes
Alkenes C=C	Bromine water is added to the sample and the mixture is shaken.	The bromine water decolourises (turns from red-brown to colourless).	Alkanes do not have a C=C bond so do not decolourise bromine water.
Halogenoalkanes R–X	First warm the sample with an aqueous solution of sodium hydroxide. This will hydrolyse the halogenoalkane and produce a halide ion. Then add nitric acid which will remove any impurities. Finally add Tollens' reagent.	A white precipitate of silver chloride shows a chloroalkane. A cream precipitate of silver bromide shows a bromoalkane. A yellow precipitate of silver iodide shows an iodoalkane.	Tollens' reagent is ammoniacal silver nitrate.
Alcohol R–OH	Add acidified potassium dichromate(VI), $K_2Cr_6O_7$ to the sample and warm the reaction mixture.	Primary alcohols are oxidised to form aldehydes and then carboxylic acids. Secondary alcohols are oxidised to form ketones. In both cases a colour change of orange to blue or green is seen.	However, aldehydes (which are not alcohols) would also be oxidised (to carboxylic acids) by acidified potassium dichromate(VI), $K_2Cr_6O_7$ so care must be taken. If the organic substance is oxidised the chromium ions in the potassium dichromate(VI), $K_2Cr_6O_7$ is reduced from +6 (orange) to +4 (blue) or +3 (green).
Aldehyde R–CHO	Add Tolles' reagent Warm the sample with Fehling's solution.	Aldehydes form a silver mirror. A change from a blue solution to a red precipitate	Ketones do not react.
Carboxylic acids R–COOH	Add an aqueous solution of sodium hydrogencarbonate to the organic sample.	Carbon dioxide gas is produced and bubbles are seen.	Carboxylic acids are weak acids and typically have a pH of 3 or 4.

16.2 Mass spectroscopy

Specification reference: 3.3.6

Modern instrumental methods

The mass spectrometer and infrared spectrometer can be used to identify compounds.

These spectrometers can be connected directly to computers, which can process enormous amounts of information at very high speed. Modern instrumental methods can also identify how much of the compound is present.

Molecular ion

The mass spectrometer can be used to measure the relative abundance of the different isotopes of an element. This information can then be used to calculate the relative atomic mass of the element.

Mass spectra can also be used to deduce the relative molecular mass of organic molecules.

The vertical axis of the mass spectrum shows the relative abundance. The horizontal axis of a mass spectrum shows the mass/charge (*m/z*) ratio of the ions reaching the detector. The ions reaching the detector have a +1 charge. These ions have lost one electron. As electrons have a negligible mass the mass/charge ratio of the heaviest ion which is called the molecular ion reveals the relative molecular mass of the organic molecule.

In Figure 1, the molecular ion has a mass / charge ratio of 75.

The relative molecular mass of propanoic acid is 75.

In the mass spectrum notice how as the organic molecules pass through the mass spectrometer some of the ions fragment into smaller pieces which may then be detected at different mass/charge ratios.

High resolution mass spectroscopy

We use high resolution mass spectroscopy to measure the mass of the molecular ion, M^+, very precisely (to lots of decimal places).

Because of the very high precision used we can determine the relative molecular mass of a compound very precisely and then use this information to deduce the identity of the substance being investigated.

Revision tip

Compared with traditional laboratory techniques, these modern methods of analysis are:

- faster
- more accurate
- more sensitive
- able to use smaller samples.

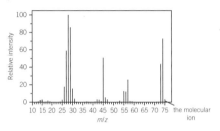

▲ **Figure 1** *The mass spectrum of propanoic acid*

Key term

Molecular ion: The ion formed when a molecule loses a single electron. The molecular ion has a *m/z* value which is equal to the molecular mass of the molecule.

Worked example

Q A sample has a relative atomic mass of 30.0688 but is it ethane or methanol?

Element	Relative atomic mass
H	1.0079
C	12.0107
O	15.9994

A Ethane, C_2H_6 has a relative atomic mass of

$$2 \times 12.0107 + 6 \times 1.0079 = 30.0688$$

While methanal, CH_2O

$$1 \times 12.0107 + 2 \times 1.0079 + 1 \times 15.9994 = 30.0259$$

So the sample is ethane, C_2H_6

Summary questions

1 What are the advantages of modern methods of analysis compared with traditional methods of analysis? *(1 mark)*

2 What is special about high-resolution mass spectroscopy? *(1 mark)*

3 What does the molecular ion tell us about an organic molecule? *(1 mark)*

Infrared spectroscopy

Different types of bonds absorb infrared radiation of slightly different wavelength. By seeing which wavelengths have been absorbed we can identify functional groups in organic molecules.

▼ **Table 1** *Identifying bonds in organic molecules*

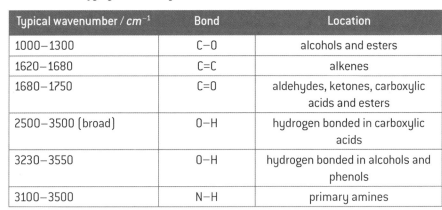

Typical wavenumber / cm^{-1}	Bond	Location
1000–1300	C–O	alcohols and esters
1620–1680	C=C	alkenes
1680–1750	C=O	aldehydes, ketones, carboxylic acids and esters
2500–3500 (broad)	O–H	hydrogen bonded in carboxylic acids
3230–3550	O–H	hydrogen bonded in alcohols and phenols
3100–3500	N–H	primary amines

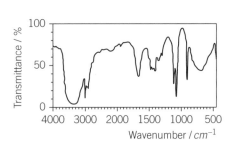

▲ **Figure 1** *Infrared spectrum of ethanol*

Ethanol

Notice that the infrared spectrum of ethanol shows a broad peak at $3400 \, cm^{-1}$ caused by the O–H bond and another at $1100 \, cm^{-1}$ caused by the C–O bond.

Identifying organic families

Alkenes

Alkenes contain C=C bonds so they will have an absorption band between 1620 and $1680 \, cm^{-1}$.

Carbonyls

Aldehydes and ketones both contain C=O bonds so they will have an absorption band between 1650 and $1750 \, cm^{-1}$.

Carboxylic acids

Carboxylic acids also contain C=O bonds so they will have an absorption band between 1680 and $1750 \, cm^{-1}$. In addition they contain an O–H bond so they will have an absorption band between 2500 and $3500 \, cm^{-1}$.

Alcohols

Alcohols contain a C–O bond so they will have an absorption band between 1000 and $1300 \, cm^{-1}$. They will also have a O–H bond so they will have a absorption band between 3230 and $3550 \, cm^{-1}$.

The fingerprint region

The area of an infrared spectrum between $400 \, cm^{-1}$ and $1500 \, cm^{-1}$ is known as the fingerprint region because it is unique for every compound. An unknown sample can be identified by comparing its infrared spectrum with a database of known infrared spectra to find a match.

Impurities

Any impurities in a sample will produce absorption bands that should not be there.

Summary questions

1 Would infrared spectroscopy allow a chemist to differentiate between a sample of ethanol and propan-1-ol? Explain your answer.
(1 mark)

2 How can you use infrared spectroscopy to show that a sample contains impurities? *(1 mark)*

3 A chemist has a sample of an organic compound which has the molecular formula of $C_3H_7O_2$. The infrared spectrum of the compound reveals an absorption band at $1700 \, cm^{-1}$ and a very broad absorption band between 2600 and $3500 \, cm^{-1}$. Deduce the identity of the organic compound. Explain your answer. *(3 marks)*

1   A student adds an organic compound containing carbon, hydrogen, and oxygen to a solution of sodium hydrogencarbonate. The student observes bubbles of a gas are produced.

   a   Name the gas produced in this reaction. *(1 mark)*

   b   Name the class of organic compound present. *(1 mark)*

2   A sample of organic compound A is analysed and is found to be a hydrocarbon.

   a   Define the term *hydrocarbon*. *(1 mark)*

   b   Describe how a student could confirm that compound A was an alkene. Include the names or any chemicals used and the results of the test in your answer. *(2 marks)*

3   An organic compound A with the molecular formula C_3H_7O was analysed using infrared spectroscopy. The infrared spectrum of compound A showed a peak at $1700\,cm^{-1}$.

   a   Which bond was responsible for the peak at $1700\,cm^{-1}$? *(1 mark)*

   b   Suggest the names of two possible identities for compound A and explain why it is not possible to be sure of the identity of compound A. *(3 marks)*

4   High resolution mass spectroscopy can be used to find the molecular mass of a parent ion to several decimal places. This can be used to identify the molecular formula of the compound being analysed.

   The table below shows the atomic masses of three elements.

Element	Accurate atomic mass
^{12}C	12.00000
^{1}H	1.007829
^{16}O	15.99491

   Compound X was analysed using high resolution mass spectroscopy and the parent ion was found to have a mass/charge ratio of 200.1049.

   It is suggested that compound X could have a molecular formula

      $C_{11}H_{20}O_3$, $C_{10}H_{16}O_4$, or $C_{11}H_4O_4$

   a   Define the term *parent ion*. *(1 mark)*

   b   Deduce the molecular formula of compound X. Explain your answer. *(2 marks)*

5   Fehling's solution can be used to differentiate between aldehydes and ketones. Which of these observations would be expected when an aldehyde is warmed with Fehling's solution?

   A   blue solution to blue precipitate

   B   decolourises

   C   blue solution to red precipitate

   D   silver mirror *(1 mark)*

6   A chemist analysed a sample of an organic compound using infrared spectroscopy.

   The infrared spectrum shows a peak at $1700\,cm^{-1}$ and a broad peak between 2500 and $3500\,cm^{-1}$. What type of organic compound had been analysed?

   A   carboxylic acid

   B   alkene

   C   alcohol

   D   aldehyde *(1 mark)*

Chapter 1

1 a $1s^2 2s^2 2p^6 3s^2$ *[1]* **b** $1s^2 2s^2 2p^6$ *[1]*

2 $\left(\dfrac{78.0}{100} \times 24\right) + \left(\dfrac{10.0}{100} \times 25\right) + \left(\dfrac{11.0}{100} \times 26\right) = 24.34$

24.3 *[2]*

3

Sub-atomic particle	Relative charge	Relative mass	
proton	+1	1	*[1]*
neutron	0	1	*[1]*
electron	−1	$\dfrac{1}{1836}$	*[1]*

4 a Isotopes have the same number of protons/ atomic number. But a different number of neutrons/mass number *[1]*

b Mass spectroscopy. *[1]*

c $\left(\dfrac{25.0}{100} \times 37\right) + \left(\dfrac{75.0}{100} \times 35\right) = 35.5$ *[2]*

36

d Cl *[1]*

5 Ar *[1]* **6** D

Chapter 2

1 a $\dfrac{2.00}{55.8} = 0.0358$ mol *[1]*

b $0.0358 \times 6.02 \times 10^{23} = 2.16 \times 10^{22}$ *[1]*

2 a The simplest whole number ratio of the atoms of each element present in the compound. *[1]*

b

P	Cl
$\dfrac{14.9}{31.0}$	$\dfrac{85.1}{35.5} = 2.40$
$\dfrac{0.481}{0.481} = 1$	$\dfrac{2.40}{0.481} = 5$

[2]

PCl_5

3 a $\dfrac{25}{1000} \times 0.1 = 0.0025$ *[2]*

b 0.0025 *[1]*

c $\dfrac{0.0025 \times 1000}{23.5} = 0.106\,mol\,dm^{-3}$ *[2]*

4 a Zinc dissolves and bubbles. *[2]*

b $\dfrac{1.50}{65.4} = 0.0229$ *[2]*

c $\dfrac{50}{1000} \times 0.2 = 0.01$ *[2]*

d Yes, the zinc is in excess. *[1]*

5 B *[1]* **6** B *[1]* **7** B *[1]*

Chapter 3

1 a Tetrahedral.

It has four bonded pairs of electrons around the central carbon atom.

It adopts this shape so the pairs of electrons are as far away from each other as possible. *[3]*

b 107°

It has three bonded pairs and one lone pair of electrons around the central nitrogen atom. Lone pairs repel more than bonded pairs. *[3]*

2 The ability of an atom to attract the electrons in a covalent bond to itself. *[2]*

3 a Metallic bonds between Mg^{2+} ions and delocalised electrons. *[4]*

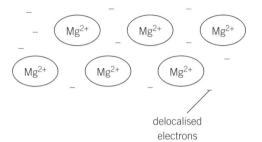

delocalised electrons

b The delocalised electrons can move. *[1]*

4 Water has hydrogen bonds between the lone pair of electrons on the oxygen atom of one molecule and the δ+ H atom of another molecule.

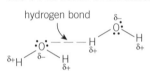

[4]

5 a

Substance	Mg	MgO
structure	giant metallic	giant ionic
type of bonding	metallic	ionic
electrical conductivity when solid	very good	poor

[3]

b The ions can move. *[1]*

c There are strong ionic bonds between the ions.

Lots of energy is required to break the bonds between these ions. *[2]*

6 a The attraction between oppositely charged ions. *[2]*

b

[1]

c There are lots of strong ionic bonds between the ions, so lots of energy is required to overcome these forces of attraction. *[1]*

7 B *[1]* **8** B *[1]*

Chapter 4

1 a Energy is required to break bonds. *[1]*

b energy in = $(4 \times 413) + 612 + 193 = 2457$

energy out = $(4 \times 413) + (2 \times 285) + 348 = 2570$

$= -113\,kJ\,mol^{-1}$ *[3]*

c It is exothermic. *[1]*

2 a Bigger surface area / reacts faster. *[1]*

b The amount of energy needed to raise the temperature of 1 g of the substance by 1 K. *[2]*

c $100 \times 4.18 \times 11.0$

$= \dfrac{4598\,J}{4.598\,K}$ or $4.598\,kJ$ *[2]*

d $\dfrac{100}{1000} \times 1.00 = 0.100\,mol$ *[1]*

e $\dfrac{4.598}{0.1} = -45.98\,kJ\,mol^{-1}$

$= -46.0\,kJ\,mol^{-1}$ *[2]*

f Heat loss

Add insulation *[2]*

3 a $C_2H_5OH(l) + 3O_2(g) \rightarrow 2CO_2(g) + 3H_2O(l)$ *[2]*

b $250 \times 4.18 \times 30$

$31350\,J$ / $31.35\,kJ$ or $31.35\,kJ$ *[2]*

c RMM of ethanol = 46.0

$\dfrac{2.00}{46.0} = 0.0435$ *[2]*

d $\dfrac{31.35}{0.0435} = -721\,kJ\,mol^{-1}$ *[2]*

4 a The enthalpy change for a reaction is independent of the route taken. *[1]*

b Lots of different hydrocarbons would be made. *[1]*

Chapter 5

1 a Energy, E *[1]*

b The peak must be shifted to the right of the original and must be lower in height than the original. *[2]*

c The particles gain energy / move faster so they collide more often and when they do collide more of the particles have the activation energy / enough energy to react. *[3]*

2 a The particles do not have enough energy / the activation energy. *[1]*

The particles are not in the correct orientation. *[1]*

b A line drawn to the left of and parallel to the original activation energy, E_a line. *[2]*

c Catalysts offer an alternative reaction pathway with a lower activation energy. *[1]*

The diagram shows how now a much greater proportion of the molecules have enough energy to react. *[1]*

3 a For a reaction to take place the particles have to collide and have enough energy to react/ activation energy. *[2]*

b Increase the temperature.

Increase the pressure/decrease the volume. *[2]*

Chapter 6

1 a

	N_2	H_2	NH_3
Amount at start	3.00	6.00	0.00
Amount at equilibrium	2.00	3.00	2.00
Concentration at equilibrium	2.00	3.00	2.00

[1]

b $K_c = \dfrac{[NH_3(g)]^2}{[N_2(g)][H_2(g)]^3}$ *[1]*

c $K_c = \dfrac{2.00^2}{2.00 \times 3.00^3} = 0.0741\,dm^6\,mol^{-2}$ *[3]*

d Decrease. The forward reaction is exothermic. *[1]*

e To the right. The right-hand side has fewer gas molecules. *[1]*

2 a Where nothing can enter or leave. *[1]*

b It is equal to the rate of backward reaction. *[1]*

c To the left. It is the endothermic direction. *[2]*

d No effect. Catalysts increase the rate of forwards and backwards reaction by the same amount. *[2]*

3 a i Increase. Particles collide more often / more particles have enough energy to react. *[1]*

ii Moves to left hand-side. This is the endothermic direction. *[1]*

b i Increase. The particles collide more often. *[1]*

ii Moves to right-hand side. This side has fewer gas molecules. *[1]*

c i Increases. More particles have enough energy to react. *[1]*

ii No effect. Catalysts increase the rate of forwards and backwards reaction by the same amount. *[1]*

Chapter 7

1 a i 3+ *[1]* **ii** 2+ *[1]*

b i iron(III) chloride *[1]* **ii** iron(II) carbonate *[1]*

2

Substance	Oxidation state of S
H_2S	−2
SO_2	+4
SO_3	+6
H_2SO_4	+6
S_8	0

[5]

3 a 0 *[1]*

b Iron is oxidised 0 to +2. *[1]*

Copper is reduced +2 to 0. *[1]*

c copper sulphate *[1]*

4 a The metal dissolves. *[1]*

Bubbles are produced. *[1]*

b Magnesium loses electrons so is oxidised. *[1]*

Hydrogen gains electrons so is reduced. *[1]*

5

	oxidation state
Mg in $MgCO_3$	+2
Cu in Cu_2O	+1
Cl in $NaClO_4$	+7

[3]

6 C *[1]* **7** B *[1]*

Chapter 8

1

Element	Bonding	Structure
Na	metallic	giant metallic lattice
Mg	metallic	giant metallic lattice
Al	metallic	giant metallic lattice
Si	covalent	giant covalent
P_4	covalent	simple molecules
S_8	covalent	simple molecules
Cl_2	covalent	simple molecules
Ar	Van der Waals	monatomic

[2]

2 a Sodium has a giant metallic structure.

Chlorine has a simple molecular structure.

The van der Waals' forces between chlorine molecules are weaker than the metallic bonds in sodium. *[3]*

b Aluminium has stronger metallic bonds than sodium.

More energy is required to break the bonds in aluminium than sodium. *[3]*

3 Sulfur *[1]*

Largest molecules / most electrons. *[1]*

Strongest van der Waals' forces between molecules. *[1]*

4 Aluminium

There is a big jump between the third and fourth ionisation energies so it must be in Group 3. *[1]*

5 Silicon has a giant covalent structure. *[1]*

There are lots of strong covalent bonds between the silicon atoms. *[1]*

A lot of energy is needed to overcome these strong forces of attraction. *[1]*

6 D *[1]*

Go Further

1 $N(g)$ $N+ (g) +e-$

2 The outer electron configuration of oxygen has a pair of electrons in one of the p orbitals. There is repulsion between these electrons so less energy is required to remove an electron.

Chapter 9

1 The first ionisation energy decreases down the group.

Although the number of protons increases, the outer electrons are further from the nucleus so the atomic radius increases and there is more shielding so the outer electron is lost more easily and the first ionisation energy decreases down the group. *[4]*

2 Mg is oxidised 0 to +2.

H is reduced +1 to 0. *[2]*

3 a Increases. *[1]* **b** Decreases. *[1]* **c** Decreases. *[1]*

4 To treat indigestion / heartburn.

It neutralises excess stomach acid. *[2]*

5 $Sr^+(g) \rightarrow Sr^{2+}(g) +e^-$

[1 mark for species]

[1 mark for state symbols]

6 a $1s^2 2s^2 2p^6 3s^2$ *[1]*

b $1s^2 2s^2 2p^6 3s^2 3p^6 4s^2$ *[1]*

c $1s^2 2s^2 2p^6 3s^2 3p^6$ *[1]*

7 B *[1]* **8** B *[1]*

Chapter 10

1 Add nitric acid. *[1]*

Then add silver nitrate. *[1]*

A white precipitate of AgCl means the halide was a chloride. *[1]*

A cream precipitate of AgBr means the halide was a bromide. *[1]*

A yellow precipitate of AgI means the halide was a iodide. *[1]*

2 a Increases. *[1]* **b** Increases. *[1]* **c** Increases. *[1]*

3 a Red / brown colour appears. *[1]*

b Cl is reduced 0 to −1. *[1]*

Br is oxidised −1 to 0. *[1]*

c chlorine *[1]*

4 D *[1]* **5** C *[1]* **6** D *[1]*

Chapter 11

1 a A compound that contains hydrogen and carbon only. *[1]*

b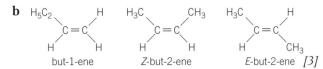

but-1-ene　　Z-but-2-ene　　E-but-2-ene *[3]*

2

2-bromopropane　　1-bromopropane *[2]*

3 C *[1]*

4 a They have the same functional group.

Each successive member differs by the addition of CH_2. *[2]*

b C_nH_{2n+2} *[1]* **c** $C_{10}H_{22}$ *[2]*

5 a The simplest whole number ratio of the atoms of each element in the compound. *[1]*

b

C	H
$\frac{83.7}{12.0} = 6.975$	$\frac{16.3}{1.0} = 16.3$
$\frac{6.975}{6.975} = 1$	$\frac{16.3}{6.975} = 2.34$

C_3H_7 *[3]*

c C_6H_{14} *[1]*

6 C *[1]*

7 a Structural isomers have the same molecular formula but a different structural formula. *[1]*

b propan-2-ol *[1]*

8 D *[1]*

Chapter 12

1 **a** A compound that contains hydrogen and carbon
only and only contains single bonds. *[2]*

b pentane *[1]*

2 **a** A substance that can be burnt to release energy. *[1]*

b $CH_4(g) + 2O_2(g) \rightarrow CO_2(g) + 2H_2O(l)$

[products = 1 mark]

[balancing = 1 mark]

c If the fuel contains sulfur when it is burnt
sulfur dioxide is formed. *[1]*

This gas reacts with water to produce acid rain. *[1]*

3 **a** zeolite / silicon dioxide and aluminium oxide *[1]*

b It has a large surface area. *[1]*

4 As the carbon chain length increases the boiling
point increases. *[1]*

The molecules get larger / have more electrons. *[1]*

The van der Waals forces are stronger. *[1]*

5 **a** $C_3H_8(g) + 5O_2(g) \rightarrow 3CO_2(g) + 4H_2O(g)$ *[2]*

b soot / carbon / carbon monoxide *[1]*

c Insufficient supply of oxygen. *[1]*

6 **a** $CH_4 + Cl_2 \rightarrow HCl + CH_3Cl$ *[1]*

b A species with an unpaired electron. *[1]*

c UV light *[1]*

d $CH_3\bullet + Cl_2 \rightarrow CH_3Cl + Cl\bullet$ *[1]*

e $2CH_3\bullet \rightarrow C_2H_6$ *[1]*

7 B *[1]* **8** A *[1]*

Chapter 13

1

1,2-dichloropropane   1,3-dichloropropane   2,2-dichloropropane

1,1-dichloropropane

[4]

2 Warm the sample with sodium hydroxide. *[1]*

Add nitric acid. *[1]*

Then add silver nitrate. *[1]*

A cream precipitate of silver bromide shows the
halogenoalkane was a bromoalkane. *[1]*

3 **a** 1-chloroethane *[1]*

1-chlorobutane *[1]*

b 1-chlorobutane *[1]*

Larger molecule / more electrons. *[1]*

Stronger van der Waals'. *[1]*

4 **a** A species with a lone pair of electrons that it can
donate to form a new covalent bond. *[1]*

b

[One mark for dipoles.]

[One mark for curly arrow from lone pair of
oxygen to δ+ C.]

[One mark for curly arrow breaking the C–Br
bond.]

[One mark for the products.]

c Nucleophilic substitution. *[2]*

5 **a**

[One mark for arrow from lone pair on oxygen
to H.]

[One arrow for curly arrow breaking C–H bond.]

[One mark for curly arrow breaking C–Cl bond.]

[One mark for products.]

b but-1-ene *[1]*

c elimination *[4]*

6 A *[1]* **7** A *[1]*

Chapter 14

1 **a** chloroethane *[1]*

b

[5]

2 a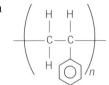

[1]

b Addition polymerisation. *[1]*

3 a A substance that can be burnt to release energy. *[1]*

b $C_3H_6(g) + 4\frac{1}{2}O_2(g) \rightarrow 3CO_2(g) + 3H_2O(g)$

[products = 1 mark]

[balancing = 1 mark]

4 a Unsaturated – has double bonds.

Hydrocarbons – only contain carbon and hydrogen atoms. *[1]*

b C_nH_{2n} *[1]* **c** C_8H_{16} *[1]*

5 a

C	H
$\frac{85.7}{12.0} = 7.14$	$\frac{14.3}{1.0} = 14.3$
$\frac{7.14}{7.14} = 1$	$\frac{14.3}{7.14} = 2$

[2]

b C_8H_{16} *[1]*

6 B *[1]* **7** D *[1]*

Chapter 15

1 a

C	H	O
$\frac{68.2}{12.0} = 5.68$	$\frac{13.6}{1.0} = 13.6$	$\frac{18.2}{16.0} = 1.14$
$\frac{5.68}{1.14} = 5$	$\frac{13.6}{1.14} = 12$	$\frac{1.14}{1.14} = 1$

[2]

b $C_5H_{12}O$ *[2]*

2 a When H_2O is removed from a substance. *[1]*

b concentrated sulfuric acid / concentrated phosphoric acid / aluminium oxide *[1]*

c pent-1-ene *[1]*

3 a C_4H_9OH or $C_4H_{10}O$ *[1]*

b Structural isomers have the same molecular formula but a different structural formula. *[1]*

c Butan-1-ol is a primary alcohol and is oxidised. *[1]*

There is a colour change orange to blue / green. *[1]*

2-methylpropan-2-ol is a tertiary alcohol and is not oxidised. *[1]*

There is no colour change. *[1]*

4 a H—C—C with H and O (aldehyde structure, ethanal) *[1]*

b H—C—C with O and OH (carboxylic acid structure) *[1]*

c H—C—C—C—H with O double bond (propanone structure) *[1]*

5 a hexanoic acid *[1]*

b i hexan-1-ol *[1]*

ii Primary. *[1]*

6 B *[1]*

Chapter 16

1 a carbon dioxide *[1]* **b** carboxylic acid *[1]*

2 a Contains carbon and hydrogen only. *[1]*

b Add bromine water and shake. *[1]*

The bromine water would decolourise if an alkene was present. *[1]*

3 a C=O *[1]*

b propanal *[1]*

propanone *[1]*

Both these compounds have the same molecular formula and both have a C=O bond which would give a peak at 1700 cm^{-1}. *[1]*

4 a A parent ion is formed when one electron is removed from the compound. *[1]*

b $C_{10}H_{16}O_4$

$(12 \times 10) + (1.007829 \times 16) + (15.99491 \times 4)$

$= 200.1049$ *[2]*

5 C *[1]* **6** A *[1]*

Answers to summary questions

1.1/1.2

1 **a** 11 protons, 12 neutrons, 11 electrons *[1]*

   b 8 protons, 8 neutrons, 10 electrons *[1]*

   c 35 electrons, 45 neutrons, 35 electrons *[1]*

   d 19 protons, 21 neutrons, 18 electrons *[1]*

2 They have the same number of electrons/same electron arrangement. *[1]*

3 Protons have a relative mass of 1 and relative charges of +1, neutrons have a relative mass of 1 and no charge. *[2]*

1.3

1 A circle/sphere. *[1]*

2 **a** $1s^2 2s^2 2p^6 3s^1$ *[1]*

   b $1s^2 2s^2 2p^6 3s^2 3p^4$ *[1]*

   c $1s^2 2s^2 2p^6 3s^2 3p^5$ *[1]*

   d $1s^2 2s^2$ *[1]*

3 **a** $1s^2 2s^2 2p^6$ *[1]*

   b $1s^2 2s^2 2p^6 3s^2 3p^6$ *[1]*

   c $1s^2 2s^2 2p^6$ *[1]*

   d $1s^2 2s^2 2p^6$ *[1]*

1.4

1 Vacuum, ionisation, acceleration, ion drift, detection, data analysis. *[1]*

2 20.179 to three significant figures the answer is 20.2. $(20.0 \times 0.909) + (21.0 \times 0.003) + (22 \times 0.088) = 20.179$ to three significant figures the answer is 20.2. *[2]*

1.5

1 $Mg^{2+}(g) \rightarrow Mg^{3+}(g) + e^-$ *[2]*

2 Although there is an increase in nuclear charge down the group, the electrons in the outer shell are further from the nucleus and there is more shielding so the outer electron is lost more easily down the group. *[2]*

3

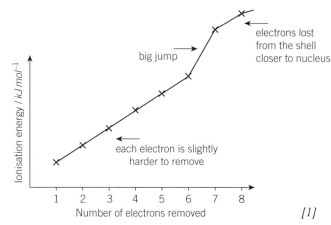

electrons lost from the shell closer to nucleus

big jump

each electron is slightly harder to remove

Ionisation energy / kJ mol⁻¹

Number of electrons removed *[1]*

2.1

1 **a** $\dfrac{5.61}{56.1} = 0.100$ moles *[1]*

   b $0.150 \times 39.1 = 5.87\,g$ *[1]*

   c $0.32 \times 23.9 = 7.6\,g$ *[1]*

2.2

1 $2.00\,mol\,dm^{-3}$ *[1]*

2 $n(HNO_3) = 0.00414$ *[1]*

   $n(NaOH) = 0.00414$ *[1]*

   $0.166\,mol\,dm^{-3}$ *[1]*

2.3

1 Vol $H_2(g) = 90.0\,cm^3$

   Ratio $H_2 : NH_3 = 3 : 2$

   Vol of $NH_3(g)$ formed $= \dfrac{90.0}{3 \times 2} = 60.0\,cm^3$ *[1]*

2 $V = \dfrac{nRT}{p} = \dfrac{2.31 \times 8.31 \times (32 + 273)}{2 \times 10^5} = 0.029\,m^3$ *[2]*

3 $n(Mg) = \dfrac{mass}{M_r} = \dfrac{0.2}{24.3} = 8.23 \times 10^{-3}$

   ratio $Mg : H_2 = 1 : 1$

   $n(H_2) = 8.23 \times 10^{-3}$

   $V = \dfrac{nRT}{p} = \dfrac{(8.23 \times 10^{-3}) \times 8.31 \times 298}{1 \times 10^5}$ *[3]*

   $= 2.04 \times 10^{-4}\,m^3$

2.4/2.5

1 $2Na(s) + Cl_2(g) \rightarrow 2NaCl(s)$ *[2]*

2 Fe $\dfrac{69.9}{55.8} = 1.25$ O $\dfrac{30.1}{16} = 1.88$

   $\dfrac{1.25}{1.25} = 1$ $\dfrac{1.88}{1.25} = 1.505$ double everything *[3]*

   2   3

   Fe_2O_3

3 $\dfrac{42.0}{14} = 3$ so the molecular formula $= C_3H_6$ *[2]*

2.6

1 % yield $= \left(\dfrac{actual\ yield}{theoretical\ yield}\right) \times 100$

   n ethene reacting $= \dfrac{2.00}{28.0} = 0.0700$ moles theoretical moles of bromoethane $= 0.0700$

   theoretical yield of bromoethane $= 0.0700 \times 108.9$

                                 $= 7.63\,g$ *[3]*

   % yield $= \left(\dfrac{5.80}{7.63}\right) \times 100 = 76.0\%$

2 atom economy $= \dfrac{relative\ mass\ of\ desired\ product}{total\ relative\ mass\ of\ products} \times 100$

   $= \dfrac{35.5 \times 2}{2(23.0 + 35.5) + 71.0 + 2} \times 100$

   $= \dfrac{71.0}{190.0} \times 100 = 37.4\%$ *[2]*

3 The percentage yield is used to compare the actual yield of a reaction with the theoretical yield of the reaction. The atom economy is the proportion of reactants that are converted into useful products. In this case the atom economy is 100% as it is an addition reaction. *[2]*

3.1

1 a NaBr *[1]*

  b Ca(OH)$_2$ *[1]*

  c MgCl$_2$ *[1]*

2 When molten the ions can move but when solid the ions cannot move. *[1]*

3.2

1 A shared pair of electrons. *[2]*

2 A shared pair of electrons but where both electrons come from the same atom. *[2]*

3 Nitrogen is a simple molecule so there are only weak forces of attraction between the molecules. *[2]*

3.3

1

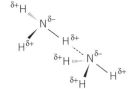

Metal ions

Delocalised electrons

The delocalised electrons can move. *[2]*

2 Sodium, magnesium, aluminium. Across the period the strength of the metallic bonding increases. *[2]*

3 In metals the layers of cations can slip over each other if a large enough force is applied. The strength of the metallic bond stops the attractions being broken completely. *[1]*

3.4

1 a $\overset{\delta+}{H}$—$\overset{\delta-}{F}$ *[1]*

  b $\overset{\delta+}{C}$—$\overset{\delta-}{Cl}$ *[1]*

  c $\overset{\delta-}{O}$—$\overset{\delta+}{N}$ *[1]*

  d $\overset{\delta+}{S}$=$\overset{\delta-}{O}$ *[1]*

2 a $\overset{\delta+}{H}$—$\overset{\delta-}{Br}$ Polar *[2]*

  b $\overset{\delta+}{H}$—$\overset{\delta-}{S}$ Polar *[2]*

  c Polar *[2]*

  d Polar *[2]*

3 a NaCl, MgCl$_2$, AlCl$_3$ *[2]*

  b NaCl, NaBr, NaI *[2]*

3.5

1

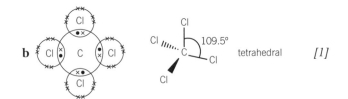 *[2]*

2 Ethane. It has more electrons therefore stronger van der Waals' forces so more energy is needed to overcome the forces of attraction. *[2]*

3.6

1

  a trigonal planar *[1]*

  b tetrahedral *[1]*

  c pyramidal *[1]*

  d bent or nonlinear *[1]*

3.7

1

Name of substance	Formula	Type of structure	Type of bonding
magnesium	Mg	giant metallic	metallic
sodium chloride	NaCl	giant ionic	ionic
chlorine	Cl$_2$	simple molecular	covalent
graphite	C	giant covalent/ macromolecular	covalent

[1]

2 Cl$_2$ – there are only weak van der Waals' forces between molecules.

H$_2$O – there are hydrogen bonds between molecules.

MgCl$_2$ – there are strong ionic bonds between the ions. *[2]*

4.1/4.2

1 Zero. Nitrogen is a diatomic gas at room temperature. *[1]*

2 $C_2H_5OH(l) + 3O_2(g) \rightarrow 2CO_2(g) + 3H_2O(g)$ *[2]*

4.3

1 Heat loss. Use a polystyrene cup or add insulation to the beaker. *[2]*

2 **a** Zinc powder has a high surface area so reacts quickly. *[1]*

 b $50.0 \times 4.18 \times (30.5 - 21.0) = 1985.5\,J$ or $1.9855\,kJ\,mol^{-1}$ *[2]*

 c moles $= \dfrac{50.0}{1000} \times 1.00 = 0.05\,mol$

 enthalpy change $= \dfrac{1.9855\,kJ}{0.05\,mol} = -39.7\,kJ\,mol^{-1}$

 The negative sign shows the reaction is exothermic. *[2]*

4.4

1 The enthalpy change of a chemical reaction is independent of the route by which the reaction is achieved and depends only on the initial and final states. *[1]*

2 $100\,kPa$ and a stated temperature of $25\,^{\circ}C$ or $298\,K$. *[2]*

3 $3C(s) + 3H_2(g) \rightarrow C_3H_6(g)$ *[2]*

4 $-(-791) + (-602) + (2 \times -33)$

 $= +123\ kJ\,mol^{-1}$ *[2]*

4.5

1 The enthalpy change when one mole of a compound is completely burnt in oxygen under standard conditions. *[1]*

2 $C_3H_6(g) + 4\frac{1}{2}O_2(g) \rightarrow 3CO_2(g) + 3H_2O(l)$ *[1]*

3 $2C(s) + 2H_2(g) \rightarrow C_2H_4(g)$

 $\downarrow\qquad\quad\downarrow\qquad\qquad\downarrow$

 Combustion products

 $+(2x-394) + (2x-286) - (-890)$

 $= -470\,kJ\,mol^{-1}$ *[3]*

4.6

1 Zero / $0\ kJ\,mol^{-1}$ *[1]*

2 Different forms of the same element that occur in the same state/phase. *[1]*

3 **a**

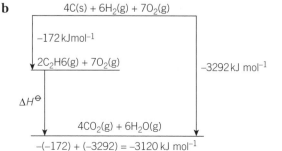

 $\Delta H^{\ominus} = -(-172) + (-3292) = -3120\,kJ\,mol^{-1}$ *[1]*

 b

$$4C(s) + 6H_2(g) + 7O_2(g)$$

$-172\,kJ\,mol^{-1}$

$2C_2H6(g) + 7O_2(g)$ $\qquad -3292\,kJ\,mol^{-1}$

$\Delta H^{\ominus}$

$4CO_2(g) + 6H_2O(g)$

$-(-172) + (-3292) = -3120\,kJ\,mol^{-1}$ *[2]*

4.7

1 The mean (average) bond enthalpy is the mean amount of energy required to break one mole of a specified type of covalent bond in a gaseous species. *[1]*

2 $H-Cl(g) \rightarrow H(g) + Cl(g)$ *[1]*

3 The mean bond enthalpy for a bond is an average value obtained from many molecules. *[1]*

4 $\dfrac{1664\,kJ\,mol^{-1}}{4} = +416\ kJ\,mol^{-1}$ *[1]*

5.1/5.2

1 It is the minimum collision energy that particles must have to react. *[1]*

2 It increases the energy of the molecules. *[1]*

3 The average energy of the particles decreases so the peak of the distribution curve moves to the left as the average energy decreases. The distribution curve becomes higher because the total number of particles remains the same. *[2]*

4 Particles collide more often and more particles have enough energy to react. *[1]*

5.3

1 A catalyst increases the rate of reaction but does not affect the yield of the reaction. *[2]*

2

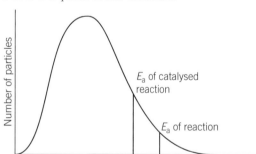

 Adding a catalyst decreases the activation energy of the reaction, a greater proportion of the particles now have enough energy to react so the rate of reaction increases. *[2]*

6.1/6.2

1 The reaction is reversible. *[1]*

2 When the conditions of a dynamic equilibrium are changed then the position of equilibrium will shift to minimise the change. *[1]*

3 It increases the rate of reaction by offering an alternative reaction pathway with a lower activation energy.

 It does not affect the position of equilibrium as the rate of forwards reaction and the rate of backwards reaction are both increased by the same amount. *[2]*

6.3

1 Where nothing can enter or leave. *[1]*

2 Iron, it increases the rate of reaction but does not affect the yield of ammonia. *[3]*

6.4

1 a $K_c = \dfrac{[NO(g)]^2\,[O_2(g)]}{[NO_2(g)]^2}$ *[1]*

b $K_c = \dfrac{[H_2(g)]\,[Cl_2(g)]}{[HCl(g)]^2}$ *[1]*

c $K_c = \dfrac{[SO_3(g)]^2}{[SO_2(g)]^2\,[O_2(g)]}$ *[1]*

2

a $\dfrac{(\cancel{mol\,dm^{-3}} \times \cancel{mol\,dm^{-3}}) \times mol\,dm^{-3}}{(\cancel{mol\,dm^{-3}} \times \cancel{mol\,dm^{-3}})} = mol\,dm^{-3}$ *[1]*

b $\dfrac{(\cancel{mol\,dm^{-3}} \times \cancel{mol\,dm^{-3}})}{(\cancel{mol\,dm^{-3}} \times \cancel{mol\,dm^{-3}})} = no\,units$ *[1]*

c $\dfrac{(\cancel{mol\,dm^{-3}} \times \cancel{mol\,dm^{-3}})}{(\cancel{mol\,dm^{-3}} \times \cancel{mol\,dm^{-3}}) \times mol\,dm^{-3}} = mol^{-1}\,dm^3$

$= dm^3\,mol^{-1}$ *[1]*

6.5

1 $\dfrac{K_c\,[HCl(g)]^2}{[H_2(g)]\,[Cl_2(g)]} = \dfrac{(0.800)^2}{0.0500 \times 0.100} = 128$ *[3]*

No units

2

	$N_2O_4(g)$	$NO_2(g)$
initial amount (mol)	0.600	0.000
amount at equilibrium (mol)	0.250	0.700
concentration at equilibrium (mol dm^{-3})	0.500	1.40

$K_c = \dfrac{[NO_2(g)]^2}{[N_2O_4(g)]}$

$= \dfrac{(1.40)^2}{0.500}$

$= 3.92\,mol\,dm^{-3}$ *[5]*

6.6

1 A decrease in pressure will shift the position of equilibrium to the left-hand side as it is the side with the more gaseous molecules. *[2]*

2 An increase in temperature will shift the position of equilibrium to the left-hand side as this is the endothermic direction. *[2]*

3 Increasing the temperature will decrease the value of the equilibrium constant, K_c as the reaction is exothermic in the forward direction. *[2]*

7.1/7.2

1 0 *[1]*

2 +6 *[1]*

3 a +5 *[1]*

b −1 *[1]*

c +7 *[1]*

7.3

1 Oxidation is the loss of electrons. Reduction is the gain of electrons. *[2]*

2 $Fe^{2+} + Zn \rightarrow Fe + Zn^{2+}$ *[1]*

3 $Zn \rightarrow Zn^{2+} + 2e^-$ *[1]*

$2H^+ + 2e^- \rightarrow H_2$ *[1]*

4 a $Mg + Cl_2 \rightarrow MgCl_2$ *[1]*

b $Mg \rightarrow Mg^{2+} + 2e^-$ *[2]*

$Cl_2 + 2e^- \rightarrow 2Cl^-$ *[2]*

8.1/8.2

a $1s^2 2s^2 2p^1$ p-block *[1]*

b $1s^2 2s^2 2p^5$ p-block *[1]*

c $1s^2 2s^2 2p^6$ p-block *[1]*

d $1s^2 2s^2 2p^6 3s^2 3p^6 3s^1$ s-block *[1]*

e $1s^2 2s^2 2p^6 3s^2 3p^6 4s^2 3d^6$ d-block *[1]*

2 Regularly recurring. *[1]*

3 a Covalent, simple molecules. *[1]*

b Metallic, giant metallic. *[1]*

4 Sulfur forms simple molecules. When sulfur melts only the weak van der Waals' forces between molecules are broken. This requires little energy and happens at low temperatures. *[1]*

Silicon forms a giant covalent structure. When silicon melts the strong covalent bonds between atoms are broken. A lot of energy is required and this only happens at high temperatures. *[1]*

8.3/8.4

1 First ionisation energy is the energy required to remove 1 electron from each atom in 1 mole of gaseous atoms forming 1 mole of ions with a single positive charge. *[2]*

2 Decrease. Each atom has one more proton in its nucleus so is able to attract its outer electron more strongly. *[2]*

3 $Mg(g) \rightarrow Mg^+(g) + e^-$ *[2]*

4 The outer electron configuration of sulfur has a pair of electrons in one of the p orbitals. There is repulsion between the electrons so less energy is needed to remove an electron from this pair so sulfur has a lower first ionisation energy than phosphorus. *[2]*

9.1

1 The solubility of Group 2 sulfates decreases down the group. *[1]*

2 The atomic radii of the elements increase down the group.

This is because down the group the atoms have another shell of electrons. *[1]*

3 The melting points generally decrease down the group.

Down the group the size of the metal ions increases so the strength of the metallic bonding decreases.

Less energy is required to overcome the forces of attraction so they melt at lower temperatures. *[1]*

10.1

1 Electronegativity is a way of measuring the attraction that a bonded atom has for the electrons in a covalent bond. *[1]*

2 Increases down the group. Down the group atoms have an extra shell of electrons and become larger. *[2]*

3 Decreases down the group. Although atoms have more protons down the group, the atoms have more shells of electrons. This means there is more shielding and less attraction between the nucleus and the electrons. In addition, down the group the atomic radius increases. This results in less attraction between the nucleus and the electrons in the covalent bond. *[2]*

10.2

1 An oxidising agent is a species which oxidises another substance by removing electrons from it. *[1]*

2 The oxidising ability of the halogens decreases down the group.

Down the group the electron which is gained is being placed into a shell which is further from the nucleus. Also the amount of shielding increases. As a result the attraction between the nucleus and the electron decreases so there oxidising ability goes down. *[3]*

3 i $Cl_2(g) + 2I^-(aq) \rightarrow 2Cl^-(aq) + I_2(aq)$ *[1]*

ii Violet *[1]*

iii The iodide, I^- ions (oxidation number −1) are oxidised to iodine, I_2 (oxidation number 0). *[2]*

10.3/4

1 a −1 *[1]*

b +5 *[1]*

2 Add silver nitrate solution and a cream precipitate would form. *[2]*

3 Down the group the halide ions become increasingly good reducing agents. Down the group the atoms get larger so it becomes easier for an electron to be lost. *[2]*

11.1

1 a Displayed formula. *[1]*

b C_4H_8 *[1]*

c $CH_3CH_2CHCH_2$ *[1]*

d CH_2 *[1]*

e *[1]*

11.2

1 a hexane *[1]*

b pentane *[1]*

c but-2-ene *[1]*

d 3-ethyl-hexane *[1]*

e 2-chlorobutane *[1]*

f 2-chloro-pentane *[1]*

11.3

1 Positional, functional, chain. *[1]*

2 *[2]*

3 *[2]*

12.1

1. A compound that only contains carbon and hydrogen atoms and only has single covalent bonds. *[2]*

2 C_nH_{2n+2} *[1]*

3 a ethane *[1]*

b propane *[1]*

c pentane *[1]*

4 a hexane *[1]*

b octane *[1]*

5 The boiling point increases as the number of carbon atoms increases. The larger the molecule is the stronger the van der Waals' forces between molecules. *[2]*

6 Propane is an alkane so is non-polar. This means it will not dissolve in polar solvents like water. *[1]*

12.2

1 When the fossilised remains of dead plants and animals are subjected to high temperatures and pressures. *[1]*

2 When burnt sulfur reacts with oxygen to make sulfur dioxide which causes acid rain. *[1]*

3 A part of the crude oil which contains alkane molecules with a similar number of carbon atoms and therefore a similar boiling point. *[1]*

4 The crude oil is heated until it vaporises. It then moves up the fractionating tower and cools. Each fraction condenses at a different point up the fractionating column where it is collected. Bitumen is collected at the bottom of the tower while refinery gases are collected at the top of the tower. *[3]*

12.3

1 a C_3H_6 *[1]*

b C_7H_{16} *[1]*

c $C_{18}H_{38}$ *[1]*

2

a [structure: hexane → butane + ethene]

$$H-C-C-C-C-C-C-H \rightarrow H-C-C-C-C-H + \quad \text{CH}_2=\text{CH}_2$$

[1]

b [structure: octane → pentane + propene]

[1]

12.4

1 A substance that can be burnt to release energy. *[1]*

2 When a substance is burnt in a large excess of oxygen. *[1]*

3 $C_4H_{10} + O_2 \rightarrow 4CO_2 + 5H_2O$ *[2]*

4 When the sulfur is burned it reacts with oxygen to form sulfur dioxide. This reacts with water to form first sulfurous acid and then sulfuric acid and this causes acid rain. *[1]*

12.5

1 A species with an unpaired electron. *[1]*

2 Free radical substitution. *[2]*

3 a UV light *[1]*

   b i $Cl_2 \rightarrow 2Cl$

   ii $Cl + C_2H_6 \rightarrow C_2H_5 + HCl$
   $C_2H_5 + Cl_2 \rightarrow C_2H_5Cl + Cl$ *[2]*

   iii $2Cl \rightarrow Cl_2$
   or $Cl + C_2H_5 \rightarrow C_2H_5Cl$
   or $2C_2H_5 \rightarrow C_4H_{10}$ *[2]*

13.1

1 a 2-bromobutane *[1]*

   b 2,2-dichloro-4-methylhexane *[1]*

2 The bond polarity of the C–X bond decreases down the group as the difference in electronegativity between the halogen and the carbon decreases down the group. *[2]*

3 The bond enthalpy of the C–X bond decreases down the group as smaller atoms attract the shared pair of electrons in the C–X bond more strongly. *[2]*

13.2

1 Nucleophiles have a lone pair of electrons which they can donate to another molecule to form a new covalent bond. *[1]*

2 Solvent. *[1]*

3 [reaction mechanism structures with $:C\equiv N^{\ominus}$ attacking and $:Cl^-$ leaving] *[4]*

13.3

1 In elimination reactions a small molecule is lost from an organic molecule. *[1]*

2 propan-2-ol *[1]*

3 [structures: pent-1-ene, Z-pent-2-ene, E-pent-2-ene] *[3]*

14.1

1 A species that can accept a pair of electrons to form a new covalent bond. *[1]*

2 Electrophilic addition. *[1]*

3 [structures: but-1-ene, Z-but-2-ene, E-but-2-ene] *[3]*

14.2

1 Electrophilic substitution. *[1]*

2
[4]

3
2-chloropropane and 1-chloropropane *[2]*

14.3

1 a poly(ethene) *[1]*

  b polychloroethene *[1]*

  c polypropene *[1]*

2
[1]

3
[1]

15.1

1 $C_nH_{2n+1}OH$. *[1]*

2 Secondary. *[1]*

3 There are hydrogen bonds between the water and methanol molecules.
[2]

15.2

1 Fermentation and hydration of ethene. *[1]*

2 Fermentation. The sugar from plants, which are living things, that can be replanted. *[1]*

3 a 100% *[1]*

  b 51% *[1]*

15.3

1 Ketone *[1]*

2 Tollen's reagent (ammoniacal silver nitrate) or Fehling's reagent. *[2]*

3 Concentrated sulfuric acid or phosphoric acid. *[1]*

4 a propanoic acid. *[1]*

  b $CH_3CH_2CH_2OH + 2[O] \rightarrow CH_3CH_2COOH + H_2O$ *[2]*

16.1

1 It would decolourise bromine water *[1]*

2 Carbon dioxide is produced so it is a carboxylic acid. *[1]*

3 Tertiary *[1]*

4 Aldehyde *[1]*

5 Warm with aqueous sodium hydroxide. Add an acidified silver nitrate. A white precipitate of AgCl is seen. *[2]*

16.2

1 Faster, more accurate, more sensitive, and smaller samples can be used. *[1]*

2 It measures the relative atomic mass of the sample very precisely / to lots of decimal places. *[1]*

3 Relative molecular mass. *[1]*

16.3

1 Although both compounds are alcohols and have the same functional group (O–H) and would show absorptions in the same areas the fingerprint region can be used to tell them apart. *[1]*

2 Absorption peaks that should not be present would indicate that a sample contains impurities. *[1]*

3 The peak at $1700\,cm^{-1}$ is caused by a C=O bond which occurs in aldehydes, ketones, carboxylic acids, and esters. *[1]*

The broad peak between 2600 and $3500\,cm^{-1}$ is caused by an O–H bond which occurs in carboxylic acids. *[1]*

So the substance must be a carboxylic acid. As it has 3 carbon atoms it is propanoic acid. *[1]*

Data

Constants

Gas constant

$R = 8.31 \, \mathrm{J\,K^{-1}\,mol^{-1}}$

Specific heat capacity

$c = 4.2 \, \mathrm{J\,g\,K^{-1}}$

Avagadro's constant

6.022×10^{23}

Infrared

Bond	Wavenumber / cm^{-1}
C—H	2850–3300
C—C	750–1100
C=C	1620–1680
C=O	1680–1750
C—O	1000–1300
O—H (alcohols)	3230–3550
O—H (acids)	2500–3000
N—H	3300–3500

Periodic Table

The Periodic Table of the elements

Key

relative atomic mass
atomic symbol
name
atomic (proton) number

1.0
H
hydrogen
1

(1)	(2)	(3)	(4)	(5)	(6)	(7)	(8)	(9)	(10)	(11)	(12)	(13)	(14)	(15)	(16)	(17)	0 (18)
																	4.0 **He** helium 2
6.9 **Li** lithium 3	9.0 **Be** beryllium 4											10.8 **B** boron 5	12.0 **C** carbon 6	14.0 **N** nitrogen 7	16.0 **O** oxygen 8	19.0 **F** fluorine 9	20.2 **Ne** neon 10
23.0 **Na** sodium 11	24.3 **Mg** magnesium 12											27.0 **Al** aluminium 13	28.1 **Si** silicon 14	31.0 **P** phosphorus 15	32.1 **S** sulfur 16	35.5 **Cl** chlorine 17	39.9 **Ar** argon 18
39.1 **K** potassium 19	40.1 **Ca** calcium 20	45.0 **Sc** scandium 21	47.9 **Ti** titanium 22	50.9 **V** vanadium 23	52.0 **Cr** chromium 24	54.9 **Mn** manganese 25	55.8 **Fe** iron 26	58.9 **Co** cobalt 27	58.7 **Ni** nickel 28	63.5 **Cu** copper 29	65.4 **Zn** zinc 30	69.7 **Ga** gallium 31	72.6 **Ge** germanium 32	74.9 **As** arsenic 33	79.0 **Se** selenium 34	79.9 **Br** bromine 35	83.8 **Kr** krypton 36
85.5 **Rb** rubidium 37	87.6 **Sr** strontium 38	88.9 **Y** yttrium 39	91.2 **Zr** zirconium 40	92.9 **Nb** niobium 41	95.9 **Mo** molybdenum 42	[98] **Tc** technetium 43	101.1 **Ru** ruthenium 44	102.9 **Rh** rhodium 45	106.4 **Pd** palladium 46	107.9 **Ag** silver 47	112.4 **Cd** cadmium 48	114.8 **In** indium 49	118.7 **Sn** tin 50	121.8 **Sb** antimony 51	127.6 **Te** tellurium 52	126.9 **I** iodine 53	131.3 **Xe** xenon 54
132.9 **Cs** caesium 55	137.3 **Ba** barium 56	138.9 **La*** lanthanum 57	178.5 **Hf** hafnium 72	180.9 **Ta** tantalum 73	183.8 **W** tungsten 74	186.2 **Re** rhenium 75	190.2 **Os** osmium 76	192.2 **Ir** iridium 77	195.1 **Pt** platinum 78	197.0 **Au** gold 79	200.6 **Hg** mercury 80	204.4 **Tl** thallium 81	207.2 **Pb** lead 82	209.0 **Bi** bismuth 83	[209] **Po** polonium 84	[210] **At** astatine 85	[222] **Rn** radon 86
[223] **Fr** francium 87	[226] **Ra** radium 88	[227] **Ac†** actinium 89	[261] **Rf** rutherfordium 104	[262] **Db** dubnium 105	[266] **Sg** seaborgium 106	[264] **Bh** bohrium 107	[277] **Hs** hassium 108	[268] **Mt** meitnerium 109	[271] **Ds** darmstadtium 110	[272] **Rg** roentgenium 111							

Elements with atomic numbers 112-116 have been reported but not fully authenticated

* 58 – 71 Lanthanides

140.1 **Ce** cerium 58	140.9 **Pr** praseodymium 59	144.2 **Nd** neodymium 60	144.9 **Pm** promethium 61	150.4 **Sm** samarium 62	152.0 **Eu** europium 63	157.3 **Gd** gadolinium 64	158.9 **Tb** terbium 65	162.5 **Dy** dysprosium 66	164.9 **Ho** holmium 67	167.3 **Er** erbium 68	168.9 **Tm** thulium 69	173.0 **Yb** ytterbium 70	175.0 **Lu** lutetium 71

† 90 – 103 Actinides

232.0 **Th** thorium 90	231.0 **Pa** protactinium 91	238.0 **U** uranium 92	237.0 **Np** neptunium 93	239.1 **Pu** plutonium 94	243.1 **Am** americium 95	247.1 **Cm** curium 96	247.1 **Bk** berkelium 97	252.1 **Cf** californium 98	[252] **Es** einsteinium 99	[257] **Fm** fermium 100	[258] **Md** mendelevium 101	[259] **No** nobelium 102	[260] **Lr** lawrencium 103